Optical Waveguides Analysis and Design

Amal Banerjee

Optical Waveguides Analysis and Design

 Springer

Amal Banerjee
Block-A
Analog Electronics
Kolkata, West Bengal, India

ISBN 978-3-030-93633-4   ISBN 978-3-030-93631-0   (eBook)
https://doi.org/10.1007/978-3-030-93631-0

This Springer imprint is published by the registered company Springer Nature Switzerland AG
The registered company address is: Gewerbestrasse 11, 6330 Cham, Switzerland

This book is dedicated to:
My late father Sivadas Banerjee
My late mother Meera Banerjee
My sister Anuradha Datta
A dear friend, mentor, and guide Dr. Andreas Gerstlauer
And the two professors who taught me the basics of dielectric optical waveguides:
Dr. John W. Keto and late Dr. A. B. Buckman

Acknowledgment

The material in this book is entirely based on the author's work experience in the telecommunication industry.

Contents

Chapter 1
Introduction and Problem Statement

1.1 Problem Statement

A waveguide is a structure or device to transfer electromagnetic signals from a source to a destination with *minimum signal energy loss*, achieved by forcing the signal to propagate in one direction only. Both an optical fiber and an optical slab waveguide operate in the near-infrared, visible frequency band of the electromagnetic spectrum. Although the underlying physics of these waveguides have been analyzed in detail in the 1950s, their full potential was realized in the 1980s, because of two key properties of optical waveguides:

- Optical waveguides support ultrahigh signal bandwidths ([1] an international team of telecommunication equipment manufacturers working with NICT, Tokyo, Japan, has demonstrated a 50 km[1] petabit/second-10^{15} bits/second optical fiber). In fact, ultrahigh bandwidth long/medium-haul (including domestic use) telecommunication links are now having all their copper wires replaced with optical fiber cables, worldwide. High bandwidths translate to concurrent transmission of very large number of signals.
- As optical waveguides are made from transparent insulating materials (glass or transparent plastic polymethyl methacrylate—PMMA), they are completely immune from electromagnetic interference, unlike copper wires.
- Optical waveguides can be fabricated from very cheap, abundant materials as glass and transparent plastics. Coupling optical signals into and out of optical fibers is straightforward. Glass slab optical waveguides can be fabricated into integrated circuits, thereby eliminating expensive, complicated integrated circuit manufacturing steps for copper interconnects between transistors.

The two interrelated approaches to analyzing, designing, and estimating performance characteristics of slab optical waveguides and fibers are geometric/ray optics and Maxwell's equations based on classical electrodynamic fields. The academic/research community is inclined towards Maxwell's equation-based approach; optical

A. Banerjee, *Optical Waveguides Analysis and Design*,
https://doi.org/10.1007/978-3-030-93631-0_1

1

fiber manufacturers [2–7] exploit concepts from both these schemes to quote performance characteristics of their optical fibers. For example, *group effective refractive index* (quoted in manufacturer-supplied datasheets of optical fibers) combines concepts from both ray optics and electromagnetic field theory. Since the potential for integrating optical slab waveguides into integrated circuits (for interconnecting transistors and other active devices) has been recognized only recently, most of the available literature (on this sub-topic) is in the form of online lecture notes, tutorials, etc. [8–25]. Traditionally, optical fibers have been analyzed in minute detail based on classical electrodynamics and Maxwell's equations in most books on advanced optics, photonics, and electronics engineering. Similarly, a vast corpus of literature [24–34] (book chapters, conference/journal papers, etc.) exists, which contains detailed analyses of the cylindrical dielectric waveguide (optical fiber).

Unfortunately, meaningful, relevant details or equations or formulas to analyze/ design/implement and evaluate performance characteristics of a fiber-optic link or optical slab waveguide cannot be extracted easily from the available literature. In addition, information about inevitable signal energy loss [35–50] in optical waveguides and how to estimate and curtail this problem is scanty. Overreliance on classical electrodynamic field theory-based analysis prevents the reader from acquiring an intuitive understanding of how light propagates through an optical waveguide. For example, low-acceptance-angle optical fiber is essential for long-haul data transmission, as it minimizes signal attenuation. The key drawbacks are listed below.

- The *fundamental or key* step to analyze and design an optical fiber link or slab waveguide segment is to determine its allowed transmission modes. This is achieved by solving a transcendental eigenvalue equation in the cylindrical or rectangular coordinate system. Such transcendental equations are derived from Maxwell's equations based on classical electrodynamic theory calculations. **Transcendental equations can only be solved iteratively and numerically, and some formulations of these equations do not converge to a solution.** *Anyone unaware of this issue would waste a lot of effort and time trying to determine and correct supposed inaccuracies in the scheme used to solve a given transcendental equation, rather than focusing on the results generated by solving the transcendental equation. This means that the author (of the textbook, etc., from where the reader obtained the transcendental equation) did not ever try to solve it numerically.*

- A miniscule minority of photonics-related literature contains incomplete actual or pseudocode to numerically solve the eigenvalue transcendental equations. The incompleteness arises from presenting the numerical solution for either the even or the odd mode for either a transverse electric (TE) or a transverse magnetic (TM) signal. *The transcendental equations for either a TE or a TM even mode are different from those for the odd mode, and need to be treated as separate cases.*

- Solving a transcendental equation involves computing and comparing large floating-point numbers, typically with some computer program. *Because computer-generated floating-point numbers can never have the same exact values, appropriate tolerances need to be applied to determine how close the matches are.* Therefore, a ready-to-use mathematical software package (Mathematica, Mathcad, MATLAB [51–53], etc.) does not generate accurate answers, as none of these packages allow the user to set such tolerances. Moreover, such mathematical software packages make simplifying assumptions in the calculation steps that introduce unwanted inaccuracies in the final results.

- For an optical fiber, the spherical coordinate system using the conventional notation (r, θ, ϕ) and after using the separation of variables technique the radial eigenvalue transcendental equation can only be solved in terms of Bessel's functions is notoriously complicated and difficult to solve, both analytically and numerically. The most effective work-around is to use the simplifying assumption called the **weak guidance** scheme. Unfortunately, literature authors spend considerable effort deriving the full-blown cylindrical coordinate system eigenvalue transcendental equation, only to point out that it cannot be used anyway. *A straightforward approach would be to simply state at the start the difficulty of solving the Bessel's function-based radial eigenvalue equation, assume the weak guidance scheme, and proceed with the analysis.*

- Apart from a few exceptions, geometric/ray optics is sidelined in the available literature, so the reader does not develop any intuitive feel for, e.g., meridional rays (rays propagating along the optical fiber axis), acceptance angle, or key V parameter. So the reader does not understand why a low-acceptance-angle optical fiber is essential for long-haul use, as it curtails signal attenuation. Neither is it clear why a single-mode fiber is preferred over multimode fiber, as a multimode fiber will have slightly different phase velocities for each propagated mode, leading to signal dispersion at the output.

1.2 Proposed Solution

Recognizing that optical waveguide (fiber, slab) is a topic in its own right, this book approaches it from both the geometric/ray optics and classical electrodynamic field and Maxwell's equations. It starts with the macro-properties of optical waveguides and drills into details step by step. Consequently, the gradual progression from geometric optics to Maxwell's equation-based formulation enables the reader to develop an intuitive understanding of how light propagates through an optical waveguide due to total internal reflection combined with constructive interference amongst the totally internally reflected light rays. Starting with the geometric/ray optics formalism automatically highlights the key concepts of V parameter, acceptance angle, beam waist, cutoff frequency, maximum possible number of modes, how a step index fiber is a special case of a graded index fiber, etc. Progressing from the top-level properties of an optical waveguide (fiber, slab) to individual propagating modes, one has to switch to electromagnetic field theoretic approach, to determine the allowed modes, group and phase velocities, and effective and group effective refractive indices. For a slab waveguide, allowed mode evaluation is accomplished by solving the appropriate (even/odd TE/TM) transcendental equation. As transcendental equations cannot be solved analytically, a set of supplied C computer language executable modules (for both the popular Linux and Windows operating systems) perform this task. As mentioned earlier, the radial eigenvalue transcendental equation (involving Bessel's functions) for an optical fiber is too complicated to be solved numerically (let alone analytically), and the *weak guidance* approximation is invoked to estimate allowed modes and their properties. *One such key optical waveguide metric is the mode chart, which in turn leads to the scheme to design an optical waveguide from predefined specifications. The slab waveguide eigenvalue transcendental equations, the calculations involved to design an optical waveguide from a set of predefined specifications, as well as all other optical waveguide analysis-related calculations (V parameter, numerical aperture, delta, etc.) are performed with a set of supplied C computer language [54] programs (for both the popular Linux and Windows operating systems). This eliminates inevitable calculation errors that would creep in if the same computations are done manually.*

For optical fibers, the **weak guidance** scheme is examined in detail to understand how it allows linearly polarized (LP) and hybrid modes (H-modes). Also, the internal signal energy loss mechanisms such as attenuation and dispersion are examined in detail, along with state-of-the-art countermeasures to tackle these issues (erbium-doped fibers). For optical slab waveguides, the distance to be traversed by the light signal is negligible compared to the distances to be traversed in an optical fiber, and consequently signal energy loss is very small compared to an optical fiber.

References

1. https://www.nict.go.jp/en/press/2020/12/18-1.html
2. https://www.corning.com/media/worldwide/coc/documents/Fiber/PI-1468-AEN.pdf
3. https://www.fujikura.co.jp/eng/resource/pdf/FutureAccess_Catalog_2017_Opticalfibersandcables.pdf
4. https://www.furukawa.co.jp/en/product/catalogue/pdf/fttx_j417.pdf
5. https://global-sei.com/fttx/product_a/opticalfibers_a/
6. https://fiber-optic-catalog.ofsoptics.com/Products/Optical-Fibers/Multimode-Optical-Fibers-3100100111
7. https://www.stl.tech/pdf/telecom-pdf/DUCT-LITE-Gel-Free-Multitube-Single-Jacket-Fibre-Optic-Cable.html
8. http://www-eng.lbl.gov/~shuman/NEXT/CURRENT_DESIGN/TP/FO/Lect4-Optical%20waveguides.pdf
9. https://courses.cit.cornell.edu/ece533/Lectures/handout8.pdf
10. https://www.fiberoptics4sale.com/blogs/wave-optics/symmetric-slab-waveguides
11. http://web.mit.edu/6.013_book/www/chapter13/13.5.html
12. http://www.phy.cuhk.edu.hk/~jfwang/PDF/Chapter%203%20waveguide.pdf
13. http://users.ntua.gr/eglytsis/IO/Slab_Waveguides_Chapter.pdf
14. Saleh Bahaa EA, Teich MC (1991) Fundamentals of photonics. John Wiley & Sons, Inc ISBNs: 0-471-83965-5 (Hardback); 0-471-2-1374-8 (Electronic)
15. http://courses.washington.edu/me557/sensors/waveguide.pdf
16. http://pongsak.ee.engr.tu.ac.th/le426/doc/slab_waveguide.pdf

17. https://www.osapublishing.org/aop/fulltext.cfm?uri=aop-1-1-58&id=176222
18. https://faculty.kfupm.edu.sa/ee/ajmal/files/Ajmal_Thesis_Chap2.pdf
19. http://www.faculty.jacobs-university.de/dknipp/c320352/lectures/3%20waveguides.pdf
20. https://link.springer.com/chapter/10.1007%2F978-1-4615-3106-7_2
21. https://link.springer.com/chapter/10.1007%2F978-0-387-73068-4_2
22. https://www.researchgate.net/publication/234932545_Analysis_of_TE_Transverse_Electric_modes_of_symmetric_slab_waveguide
23. http://www.m-hikari.com/astp/astp2012/astp25-28-2012/ramzaASTP25-28-2012.pdf
24. http://opticaapplicata.pwr.edu.pl/files/pdf/2015/no3/optappl_4503p405.pdf
25. https://arxiv.org/pdf/1511.03479.pdf
26. https://www.osapublishing.org/josa/abstract.cfm?uri=josa-51-5-491
27. http://users.ntua.gr/eglytsis/IO/Cylindrical_Waveguides_p.pdf
28. Kapoor A, Singh GS (2000) Mode classification in cylindrical dielectric waveguides. J Lightwave Technol 18(6)
29. https://www.sciencedirect.com/topics/physics-and-astronomy/dielectric-waveguides
30. https://link.springer.com/chapter/10.1007/978-0-387-49799-0_5
31. https://www.researchgate.net/publication/232986168_Single-mode_Anisotropic_Cylindrical_Dielectric_Waveguides
32. Yu E, Smolkin, Valovik DV (2015) Guided electromagnetic waves propagating in a two-layer cylindrical dielectric waveguide with inhomogeneous. Nonlinear Permittivity Adv Math Phys:614976. https://doi.org/10.1155/2015/614976
33. Novotny L, Hafner C (1994) Light propagation in a cylindrical waveguide with complex, metallic, dielectric function. Phys Rev E50:4094
34. Bruno WM, Bridfes WB (1988) Flexible dielectric waveguides with powder cores. IEEE Trans Microwave Theory Tech 36(5)
35. https://www.ad-net.com.tw/optical-power-loss-attenuation-in-fiber-access-types-values-and-sources/
36. https://www.ppc-online.com/blog/understanding-optical-loss-in-fiber-networks-and-how-to-tackle-it
37. https://www.juniper.net/documentation/en_US/release-independent/junos/topics/concept/fiber-optic-cable-signal-loss-attenuation-dispersion-understanding.html
38. https://www.fiberoptics4sale.com/blogs/archive-posts/95048006-optical-fiber-loss-and-attenuation
39. https://www.sciencedirect.com/topics/engineering/fiber-loss
40. https://community.fs.com/blog/understanding-loss-in-fiber-optic.html
41. https://www.elprocus.com/attenuation-in-optical-fibre-types-and-its-causes/
42. https://www.cablesys.com/updates/fiber-loss-in-fiber-optic-performance/
43. https://www.thefoa.org/tech/ref/testing/test/dB.html
44. http://opti500.cian-erc.org/opti500/pdf/sm/Module3%20Optical%20Attenuation.pdf
45. https://wiki.metropolia.fi/display/Physics/Attenuation+in+Fiber+Optics
46. http://www.dsod.p.lodz.pl/materials/OPT04_IFE.pdf
47. http://electron9.phys.utk.edu/optics421/modules/m8/optical_fiber_measurements.htm
48. https://rmd.ac.in/dept/ece/Supporting_Online_%20Materials/7/OCN/unit2.pdf
49. https://www.rp-photonics.com/passive_fiber_optics7.html
50. https://link.springer.com/chapter/10.1007%2F978-1-4757-1177-6_19
51. https://www.wolfram.com/mathematica/
52. https://www.mathworks.com/products/matlab.html
53. https://www.mathcad.com/en
54. The all-time classic book on C computer language, written by the creators: https://www.amazon.in/Programming-Language-Prentice-Hall-Software/dp/0131103628

Chapter 2
Transparent Dielectric Optical Waveguide Fundamentals

2.1 Electromagnetic Wave Propagation Modes Maxwell's Equations TEM, TE, and TM

Geometric/ray optics [1–28] assumes that light propagates as plane waves through a homogeneous medium, with the "ray" indicating the direction of propagation of the electromagnetic wave, since light is a form of electromagnetic wave.

To understand this, Maxwell's equations of classical electrodynamics must be examined, starting with propagation in free space. *For the remainder of this section, all electric and magnetic field, electric and magnetic field densities, etc. are vector quantities:*

$$\nabla \times E = \frac{-\mu_0 \partial H}{\partial t}, \quad \nabla \times H = \frac{\epsilon_0 \partial E}{\partial t}, \quad \nabla H = 0 \quad \text{and} \quad \nabla E = 0 \quad (2.1a)$$

where $\epsilon_0 = 8.854 \times 10^{-12}$ and $\mu_0 = 12.57 \times 10^{-7}$ are, respectively, the permittivity and permeability of free space, and E and H are the electric and magnetic fields. These equations can be extended to conducting (with free mobile charges) medium in which case the electric and magnetic fields are replaced by the electric (displacement currents because of mobile charges) and magnetic flux densities $D = \epsilon_0 E + P$ and $B = \mu_0(H + M)$ where M and P are, respectively, the magnetization and polarization densities. The polarization density is the sum total of the polarization induced in the medium by the electric field, and magnetization density is the sum total of the magnetization induced by the magnetic field. *In a homogeneous medium,*

The original version of this chapter was revised. The correction to this chapter is available at https://doi.org/10.1007/978-3-030-93631-0_4

Supplementary Information The online version contains supplementary material available at (https://doi.org/10.1007/978-3-030-93631-0_2).

A. Banerjee, *Optical Waveguides Analysis and Design*, https://doi.org/10.1007/978-3-030-93631-0_2

all components of the electric and magnetic fields must be continuous. In the absence of free charges and currents at the boundary of two dielectric media, the tangential components of E and H and the normal components of B and D are continuous. The power carried by an electromagnetic wave is $S = E \times H$—Poynting vector.

Maxwell's equations in a dielectric medium can be formulated and understood taking into account the following key material properties.

- For a *linear* dielectric medium, the polarization vector is linearly related to the electric field vector.
- For a *nondispersive* dielectric medium, the response of the polarization vector is *instantaneous* to the electric field vector.
- A dielectric medium is *homogeneous* if the polarization vector is related to the electric vector *independent* of the position.
- A dielectric medium is *isotropic* if the relation between the polarization and electric vector is *independent of the direction* of the electric vector.
- A dielectric medium is *spatially isotropic* if the polarization at a location depends on the electric field at that location *only*.

Combining the above for a linear, nondispersive, and isotropic medium, the polarization vector is related to the electric vector as

$P = \epsilon_0 \xi \, E$ where ξ is a scalar called the electric susceptibility. Maxwell's equations in this case become $\nabla \times H = \epsilon \frac{\partial E}{\partial t}$, $\nabla \times E = -\mu_0 \frac{\partial H}{\partial t}$, $\nabla E = 0$, and

$$\nabla H = 0 \quad \text{where} \quad \epsilon = \epsilon_0(1 + \xi). \tag{2.1b}$$

Both the electric and magnetic fields satisfy the wave equation:

$$\nabla^2 u - \frac{\partial^2 u}{c^2 \partial t^2} = 0 \quad \text{where} \quad c = \frac{1}{\sqrt{\epsilon \mu_0}} \tag{2.2a}$$

2.1.1 Inhomogeneous Medium

A graded index material is inhomogeneous, but linear, nondispersive, and isotropic. Then both ϵ, ξ are functions of position. The corresponding wave equation for E is $\nabla^2 E - \frac{\partial^2 E}{c^2(r)\partial t^2} = 0$ where $c(r) = \frac{c_0}{n(r)}$ and $n(r) = \sqrt{\frac{\epsilon(r)}{\epsilon_0}}$. The wave equation, after some manipulation, gives

$$\nabla^2 E - \frac{\partial^2 E}{c^2(r)\partial t^2} + \nabla \left(\frac{\nabla \epsilon . \, E}{\epsilon} \right) = 0 \tag{2.3}$$

2.1.2 Nonisotropic Medium

When the polarization vector depends on the direction of the electric field vector E, if the medium is linear, nondispersive, and homogeneous, each component of the polarization vector depends on the three components of the electric vector.

2.1.3   Dispersive Medium

Dispersive medium is the most interesting case. The electric field vector E of the passing electromagnetic wave forces the bound electrons of the dielectric medium atoms to oscillate, causing polarizations. There is a finite time interval between the electromagnetic wave arrival at an atom and for bound electrons to respond and align with the electromagnetic field. If this interval is very short, the effect is termed instantaneous and the medium is basically nondispersive. Typically, for analysis purposes, the medium is dispersive, but linear, homogeneous, and isotropic. *It is dispersive because by aligning the bound electrons, the passing electromagnetic wave loses a very small percentage of its energy.*

The dynamic relationship between the driving electromagnetic field and the responding bound electrons is governed by the linear differential equation for a driven harmonic oscillator. A linear system is characterized by its response to the impulse function (Dirac delta function). An electromagnetic impulse $\delta(t)$ generates a time-dispersive polarization response as

$P(t) = \epsilon_0 \int\limits_{-\infty}^{\infty} x(t - t')E(t')dt'$ where $x(t)$ is a scalar function of time. A linear system can also be analyzed using its transfer function.

2.1.4   Nonlinear Medium

In a nonlinear medium, the relation between the polarization vector and electric field vector is nonlinear. Specifically, if the medium is homogeneous, nondispersive, and isotropic, then $P = \Psi(E)$ at each position and time instance. Although Maxwell's equations cannot be used to analyze electromagnetic wave propagation in a nonlinear medium, they can be used to obtain a partial differential equation to achieve this goal, starting with $\nabla \times H = \frac{\partial D}{\partial t}$, along with $B = \mu_0 H$ and the curl operator, $\nabla \times (\nabla \times E) = -\mu_0 \frac{\partial^2 D}{\partial t^2}$. Now $D = \epsilon_0 E + P$ and exploiting the relation $\nabla \times (\nabla \times E) = \nabla(\nabla . E) - \nabla^2 E$ the following partial differential equation is obtained:

$$\nabla(\nabla . E) - \nabla^2 E = -\epsilon_0 \mu_0 \frac{\partial^2 E}{\partial t^2} - \epsilon_0 \frac{\partial^2 P}{\partial t^2} \tag{2.4}$$

This equation, after some more manipulation, gives

$$\nabla^2 E - \frac{\partial^2 E}{c_0^2 \partial t^2} = \frac{\epsilon_0 \partial^2 \Psi}{\partial t^2} \tag{2.5}$$

For a plane electromagnetic wave, **both the electric and magnetic field vectors are perpendicular(transverse) to the propagation (longitudinal) direction.** So, plane wave propagation is termed as transverse electric magnetic (TEM). However, no waveguide supports pure TEM waves, rather either transverse electric (TE) or transverse magnetic (TM) waves only. **For a TE electromagnetic wave, the magnetic field vector is parallel to the direction of propagation, whereas for a TM wave, the electric field vector is parallel to the direction of propagation.** "Transverse" is any direction perpendicular to the direction of propagation. Using simplifying assumptions, one can reason about "quasi-TEM" waveguide propagation modes.

2.1.5 Electromagnetic Wave Polarization

Polarization of an electromagnetic wave (e.g., light) is determined by how the direction of the electric field vector varies with time. The components of electric field vector of a monochromatic electromagnetic wave vary sinusoidally with time, with different amplitudes and phases, such that at each position $\vec{r}$ the endpoint of the electric field vector $\vec{E}\left(\vec{r},\ t\right)$ traces an ellipse. The plane, shape, and orientation of the ellipse vary with position. For an isotropic medium, the ellipse is the same everywhere, and the electromagnetic field is elliptically polarized. The orientation and ellipticity of the ellipse are different for different types of electromagnetic waves and so the electromagnetic wave can be linearly or circularly polarized.

- Polarization determines the amount of light reflected at the boundary between two transparent (to electromagnetic waves) media.
- Polarization influences electromagnetic wave scattering absorption and scattering by different transparent media.
- Refractive index of anisotropic medium depends on the polarization; therefore electromagnetic waves of different polarizations move with different phase velocities and phases in that anisotropic medium.
- Optically active materials can rotate the polarization vector of linearly polarized electromagnetic waves.

A monochromatic light wave traveling in the z direction, with velocity c, has an electric field vector $\vec{E}\left(\vec{r},\ t\right) = \Re\left(\vec{A}\ e^{j3\$\ pi\left(t-\frac{z}{c}\right)}\right)$ where $\vec{A} = A_x\widehat{x} + A_y\widehat{y}$ is the complex envelope. The components of the electric field vector are $E_x = a_x \cos\left(2\pi\ f\left(t-\frac{z}{c}\right) + \phi_x\right)$ where ϕ_x is the phase angle and $E_y = a_x \cos\left(2\pi\ f\left(t-\frac{z}{c}\right) + \phi_y\right)$.

These two equations, after some manipulation, give

$$\frac{E_x^2}{a_x^2} + \frac{E_y^2}{a_x^2} - 2\cos\left(\phi_x - \phi_y\right)\frac{E_x\,E_y}{a_x\,a_y} = \sin\left(\left(\phi_x - \phi_y\right)\right)^2 \qquad (2.6)$$

Fig. 2.1 Polarization ellipse and helical path traced by the tip of the electric field vector as electromagnetic wave propagates along the z-axis

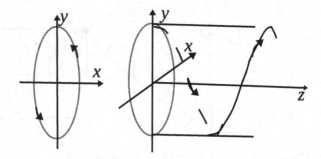

Fig. 2.2 Linearly polarized electromagnetic wave

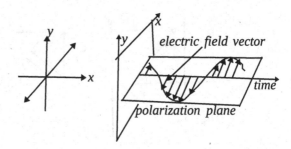

This equation implies that at a fixed value of z, the tip of the electric field vector traces an ellipse in the x–y plane; that is, at a fixed time t, the locus of the electric field vector traces a helical trajectory in space, on the surface of an elliptical cylinder (Fig. 2.1). The shape of the ellipse depends on the ratio of the electric field vector component amplitudes and the phase difference, $\frac{a_x}{a_y}$, $\phi =$ $\phi_1 - \phi_2$. The size of the ellipse depends on the intensity of the wave $I = \frac{\left(a_x^2 + a_y^2\right)}{2material_{impedance}}$.

There are several common polarizations: linear, planar, and circular (left, right, Figs. 2.2, 2.3, and 2.4).

The matrix form and Jones vector are very convenient schemes to represent the polarization of an electromagnetic (light) wave. An electromagnetic wave (light) traveling in the z direction is identified by the complex envelopes in the x and y directions as $A_x = a_x e^{j\phi}$ and $A_y = a_y e^{j\phi_y}$ and it is convenient to represent this information in vector (Jones) $\vec{J} = \begin{bmatrix} A_x & A_y \end{bmatrix}$. The Jones vectors for some common cases are:

- Linearly polarized (along the x-axis) wave $\vec{J} = \begin{bmatrix} 1 & 0 \end{bmatrix}$
- Linearly polarized (making angle θ with the x-axis) wave $\vec{J} = \begin{bmatrix} \cos(\theta) & \sin(\theta) \end{bmatrix}$
- Right circularly polarized light $\vec{J} = \frac{1}{\sqrt{2}}[1 j]$
- Left circularly polarized light $\vec{J} = \frac{1}{\sqrt{2}}[1 - j]$

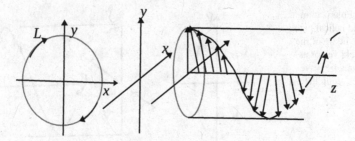

Fig. 2.3 Left circularly polarized electromagnetic wave

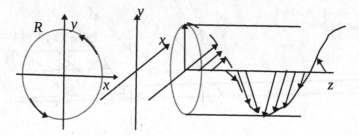

Fig. 2.4 Right circularly polarized electromagnetic wave

Two polarized electromagnetic waves are orthogonal if the inner product of their respective Jones vectors is zero: $\left(\vec{J}_1 \cdot \vec{J}_2\right) = A_{1x}A_{2x}^P + A_{2y}A_{2y}^P = 0$.

The Jones matrix scheme allows representation of an arbitrary polarization as a superposition of orthogonal Jones vectors $\left(\vec{J}_1, \vec{J}_2\right)$, so that any arbitrary Jones vector can be represented as $\vec{J} = \alpha_1 \vec{J}_1 + \alpha_2 \vec{J}_2$, where α_1, α_2 are arbitrary weights. In most common cases, $\alpha_i = \vec{J} \cdot \vec{J}_i$ $i = 1, 2$. For example, for left and right circular polarizations with Jones bases $\vec{J} = \frac{1}{\sqrt{2}}[1j]$ and $\vec{J} = \frac{1}{\sqrt{2}}[1 - j]$; the expansion coefficients are $\alpha_1 = \frac{1}{\sqrt{2}}\left(A_x - j\ A_y\right)$ and $\alpha_2 = \frac{1}{\sqrt{2}}\left(A_x + j\ A_y\right)$. Extending this concept further, if an optical system alters the polarization of an incident wave, then the retransformation can be represented as a matrix:

$$A_{2x} = T_{11}\ A_{1x} + T_{22}\ A_{1y} \text{and} A_{2y} = T_{21}\ A_{2x} + T_{22}\ A_{2y}$$

where T_{ij} are the elements of the transformation matrix. Expressed in terms of Jones vectors, the Jones vector for the wave after polarization is $\vec{J}_2 = T\ \vec{J}_1$, where T is the Jones matrix for the transformation. Therefore, the Jones matrix for linear polarization along the x-axis is $T = [1000]$ and the Jones matrix for a wave retarder (with fast axis along the x-axis) is $T = [100\ e^{-jT}]$ and that for a polarization rotator is $T = [\cos(\theta) - \sin(\theta)\sin(\theta)\cos(\theta)]$. These Jones matrices can be cascaded, and the Jones matrix for an entire optical system can be created by multiplying the individual Jones matrices for the components of the system.

Fig. 2.5 Reflection and refraction at the boundary of two transparent dielectric media

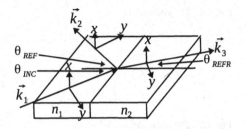

With the Jones matrix concept in mmd, reflection and refraction at the boundary of two transparent dielectric media with refractive indices n_1, n_2 $n_1 < n_2$ are examined as in Fig. 2.5. At the interface, $\theta_{INC} = \theta_{REF}$ $n_1 \sin(\theta_{INC}) = n_2 \sin(\theta_{REFR})$ (Snell's law). To relate the polarizations, the three Jones vectors for the incident, reflected, and refracted waves are $\vec{J}_1 = \begin{bmatrix} A_{1x} & A_{1y} \end{bmatrix}$, $\vec{J}_2 = \begin{bmatrix} A_{2x} & A_{2y} \end{bmatrix}$, and $\vec{J}_3 = \begin{bmatrix} A_{3x} & A_{3y} \end{bmatrix}$, respectively. Using the transmission and reflection matrices, $\vec{J}_3 = \overline{T}\vec{J}_1$ and $\vec{J}_2 = \overline{R}\,\vec{J}_1$ where R and T are the reflection and transmission matrices, respectively. The elements of these two matrices are determined by applying the boundary conditions that the tangential components of the electric field and magnetic flux $\vec{E}$ $\vec{H}$ are continuous, and the normal components of the displacement current and magnetic flux density $\vec{D}$ $\vec{B}$ are continuous. The magnetic field vectors for each of the incident, reflected, and refracted cases are orthogonal to the electric field vectors, and so their magnitudes are related by the ratio of the impedances of the respective media $\frac{\eta_0}{n_1}, \frac{\eta_0}{n_1}, \frac{\eta_0}{n_2}$ $\eta_0 = \sqrt{\epsilon_0 \mu_0}$. The subsequent mathematical manipulations are simplified by observing that the waves for each of the incident, reflected, and refracted waves are linearly polarized. The x-axis polarized mode is called the *transverse electric (TE) or orthogonal polarization* since the electric field vectors are orthogonal to the plane of incidence, and the y-axis polarized magnetic field vectors are called the *transverse magnetic (TM) or parallel. Based on this information, the transmission and reflection matrices are $T = [t_x \ 0 \ t_y \ 0]$ and $R = [r_x \ 0 \ r_y \ 0]$.*

Then the reflected and refracted electric vector components are $E_{2x} = r_x E_{1x}$, $E_{2y} = r_y E_{1x}$ $E_{3x} = t_x E_{1x}$ $E_{2y} = t_y E_{1y}$.

The matrix elements are obtained by applying the above expressions to the boundary conditions and are given by Fresnel's equations:

$$r_x = \frac{n_1 \cos(\theta_{INC}) - n_2 \cos(\theta_{REFR})}{n_1 \cos(\theta_{INC}) + n_2 \cos(\theta_{REFR})} \quad r_y = 1 + r_x \qquad (2.7a)$$

$$t_x = \frac{n_1 \cos(\theta_{INC}) - n_2 \cos(\theta_{REFR})}{n_1 \cos(\theta_{INC}) + n_2 \cos(\theta_{REFR})} \quad t_y = \frac{n_1}{n_2}(1 + t_x) \qquad (2.7b)$$

Fig. 2.6 TE polarization

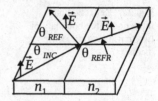

Fig. 2.7 TM polarization

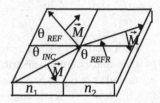

With knowledge of n_1, n_2, θ_{INC} the value of the angle of refraction is calculated using $\theta_{INC} = \sqrt{\left(1 - \left(\frac{n_1\,\theta_{INC}}{n_2}\right)^2\right)}$.

For TE polarization, the reflection coefficient is given in Eq. (2.7a) (Fig. 2.6). Two types of reflection occur: external and internal. For external reflection, $n_1 < n_2$ and the reflection coefficient is $r_x < 0$, *real* and the phase shift is $\phi = \pi$ $|r_x| = \frac{n_2-n_1}{n_1+n_2}$ *normal incidence* and increases to 1 by grazing incidence. For internal reflection, $n_1 > n_2$ for small angles of incidence, $r_x > 0$, *real*. Its magnitude is $|r_x| = \frac{n_1-n_2}{n_1+n_2}$. When the angle of incidence becomes greater than the critical angle, the reflection coefficient becomes unity, ensuring that all the electromagnetic radiation(light) is reflected back into the same medium from which it was incident on the interface. In this case, the phase shift is $\tan\left(\frac{\phi_x}{2}\right) = \frac{\sqrt{((\sin(\theta_{INC}))^2 - (\sin(\theta_{CRITICAL}))^2)}}{\cos(\theta_{INC})}$

For TM polarization, for internal reflection (Fig. 2.7), $\tan\left(\frac{\phi_y}{2}\right) = \frac{\sqrt{((\sin(\theta_{INC}))^2 - (\sin(\theta_{CRITICAL}))^2)}}{\cos(\theta_{INC})\,(\sin(\theta_{CRITICAL}))^2}$ and for external reflection, $n_1 < n_2$, and the reflection coefficient decreases from $\frac{n_2-n_1}{n_2+n_1}$ and becomes zero when $\theta_{INC} = \theta_{BREWSTER} = \arctan\left(\frac{n_2}{n_1}\right)$.

2.1.6 Absorption and Attenuation of Electromagnetic Waves

As an electromagnetic wave (light) propagates through a dielectric medium (e.g., glass) it loses energy due to absorption by the medium and resultant attenuation. Therefore, all transparent dielectric media are classified according to the sub-band of the far infrared to ultraviolet frequency band where they permit the electromagnetic wave to pass through without absorption. All transparent dielectric media have couple permittivity and susceptibility.

$\epsilon = \epsilon_R + j\ \epsilon_I$ and $\xi = \xi_R + j\ \xi_I$, and although Helmholtz's equation holds, $k = \omega\sqrt{\epsilon\ \mu_0} = k_0\sqrt{1 + \xi_R + j\ \xi_I}$. This means that for a plane wave traveling in such a dielectric medium, both the amplitude and phase are complex.

To comment on the complex susceptibility to the propagation (rate at which the wave's phase changes with respect to the direction of travel—z) and attenuation (positive/negative) constants (β, α *real*), $\beta - \frac{j\ \alpha}{2} = nk_0 - \frac{j\ \alpha}{2} = k_0\sqrt{1 + \xi_R + j\ \xi_I}$. So, the refractive index of the transparent dielectric medium and the absorption coefficient are related to the real and imaginary parts of the susceptibility as $n - \frac{j\ \alpha}{2 k_0} = \sqrt{1 + \xi_R + j\ \xi_I}$. For a weakly absorbing medium when both the real and imaginary parts of the complex susceptibility are much smaller than 1, $n = 1 + \frac{\xi}{2}$ $\alpha = -k_0\xi$.

Transparent dielectric media, for which the susceptibility, refractive index, and phase velocity of electromagnetic waves (light) are frequency/wavelength dependent, are called **dispersive**. So, these are optical components, as prisms refract polychromatic light at different angles, because the refractive index for each frequency is different. More importantly, as the phase velocities of each frequency of the polychromatic light are different, each of those frequency components takes different time so that a polychromatic light pulse is dispersed in time and if the transparent dielectric medium is long (long-haul optical fiber) the output pulse would appear broadened (Fig. 2.8).

Dispersion is measured using several methods. One of them is the ∇ number defined as $\nabla = \frac{n_D - 1}{n_F + n_C}$ where n_F, n_D, n_C are, respectively, the refractive indices of blue, yellow, and red.

Another dispersion measurement scheme at a particular frequency λ_{OBS} is $\frac{dn}{d\lambda_{OBS}}$. As the angle of refraction is a function of frequency (refractive index is a function of frequency) the dispersion angle θ_d varies as a function of the selected wavelength (at which measurement is made) as $\frac{d\ \theta_d}{d\ \lambda_{OBS}} = \frac{d\ \theta_d}{dn}\frac{dn}{d\ \lambda_{OBS}}$, i.e., product of material dispersion and a factor that depends on the geometry and refractive index. The first and second derivatives $\frac{d\ n}{d\ \lambda_{OBS}}$, $\frac{d^n}{d\lambda_{OBS}^2}$ control the dispersion of an electromagnetic pulse through dispersive medium. When a pulse of free space wavelength λ_0 travels with a group velocity $c_{GRP} = \frac{c}{n - \lambda_0\frac{dn}{d\ \lambda_0}}$ the pulse is broadened per unit distance at a rate $|D_\lambda|\ \sigma_\lambda$ where σ_λ is the spectral width of the pulse and $D_\lambda = \frac{-c}{\lambda_0}\frac{d^2n}{d\ \lambda_0^2}$ is the

Fig. 2.8 General signal dispersion

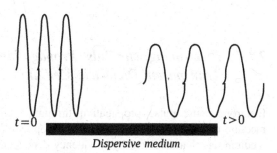

Dispersive medium

dispersion coefficient. *This dispersion measure is used as a performance metric for long-haul optical fibers.*

Dispersion and absorption in transparent dielectric medium are interrelated. A medium with frequency/wavelength-dependent refractive index will also be absorptive and the absorption coefficient will be frequency dependent. **This interdependency is because of the interrelationship between the frequency-dependent real and imaginary parts of the complex susceptibility, expressed as Kramers–Kronig relations:**

$$\xi_R(f) = \frac{2}{\pi} \frac{\int s\, \xi_I(s) d\, s}{s^2 - f^2} \text{ and } \xi_I(f) = \frac{2}{\pi} \frac{\int s\, \xi_R(s) d\, s}{s^2 - f^2} \tag{2.8}$$

So if either the real or the imaginary part of the complex susceptibility is known for all frequencies, the other can be determined. That information cam be used to compute the frequency-dependent refractive index, absorption coefficient, etc.

A resonant transparent dielectric medium is one for which the polarization and the electric field are linked via the damped harmonic oscillator equation:

$$\frac{d^2 P}{d\, t^2} + \frac{\sigma}{d\,t} \frac{d\,P}{dt} + \omega_0^2 P = \omega_0^2\, \epsilon_0 \xi_0\, E$$

where σ, ϵ_0, ξ_0 are constants. Using $E(t) = \Re\left(\vec{E} e^{j\,\$\,omega\,t}\right)$ and $P(t) = \Re\left(\vec{P} e^{j\,\$\,omega\,t}\right)$, the above equation reduces to $(-\omega^2 + j\,\sigma\,\omega + \omega_0^2)P = \omega_0^2\,\epsilon_0\xi_0 E$. The expression for P is manipulated and this gives the complex susceptibility expression as $\xi(f) = \frac{\xi_0 f_0}{2((\,f-f_0) + \frac{j\,\Delta\,f}{2})}$. This expression, after some manipulation, gives the real and imaginary parts of the complex susceptibility as $\xi(f) = \frac{2(\,f-f_0)\xi_I}{\Delta\,f}$ and $\xi_I(f) = \frac{-\xi_0 f_0 \Delta f}{4\left((\,f-f_0)^2 + \left(\frac{\Delta\,f}{2}\right)^2\right)}$. Now using the definitions of the refractive index and absorption coefficients their dependency on the complex susceptibility is

$$n(f) = n_0 + \frac{\xi_R(f)}{2\,n_9} \quad \text{and} \quad \alpha(f) = -\left(\frac{2\pi\,f}{n_0\,c}\right) \xi(f).$$

2.1.7 Electromagnetic Pulse Propagation Through Dispersive Transparent Dielectric Medium

Electromagnetic pulse propagation through a dispersive transparent dielectric medium describes how light traverses an optical fiber or slab waveguide. Such a medium is characterized by frequency-dependent refractive index, absorption coefficient, and phase velocity. So, the different frequency constituents of light pulse have different transit times starting from optical fiber entry point to the exit. Each of these frequency constituents undergoes different attenuation, so that the exiting pulse is flattened with respect to the entering pulse. This analysis assumes that the medium is linear, isotropic, and homogeneous, and the frequency-dependent propagation constant is $\beta(f) = \frac{2\pi \; fn(f)}{c}$, and the complex wave function is $U(z, \; t) = A(z, \; t)e^{\,j(2\pi f_0 - \beta_0 z)}$ where f_0, $\beta(f_0)$ are, respectively, the central frequency and propagation at the central frequency. The complex amplitude $A(z, t)$ varies very slowly compared to the central frequency. The input and output complex envelopes can be regarded as the input and output of a linear system. The complex envelopes at entry are $A(0, \; t) \; e^{2\pi j \, f^{P} t}$ and $f = f^{P} + f_0$, the wave is monochromatic, and its complex wave function is $U(z, \; t) = U(0, \; t)e^{\frac{-\alpha(\; f^{P}+f_0)}{2}-j\beta(\; f^{P}+f_0)z}$. The complex amplitude is written as $A(z, \; f^{P}) = A(0, f^{P})e^{\frac{-\alpha(\; f^{P}+f_0)z}{2}-j[\beta(\; f^{P}+f_0)]z}$ where the term in the exponent is the transfer function of the linear system. The frequency and time domain expressions (Fourier transforms of each other) for the complex envelopes are given as

$$A(z, \; f^{P}) = \int A(z, \; t)e^{-2j\pi f^{P} t}dt \; -\infty \leqslant t \leqslant \infty$$

and

$$A(z, \; t) = \int A(z, \; f^{P})e^{2j\pi f^{P} t}d f^{P} \; -\infty \leqslant f^{P} \leqslant \infty$$

In the slowly varying envelope approximation, the complex amplitude varies very slowly varying the function of f^{P} and $\Delta f \ll f_0$. In addition, the attenuation coefficient is approximately constant, and the propagation constant varies very slowly with respect to the frequency, so that it can be approximated with a Taylor's series expansion with the first three terms, i.e., $\beta(\; f_0 + f^{P}) = \beta(\; f_0) + f^{P}\frac{d\beta}{df} + (\; f^{P})^2 \frac{d^2\beta}{df^2}$. Using this expansion, the approximate expression for the transfer function becomes $e^{\frac{\alpha}{2}z} \; e^{-j2\pi f^{P}\frac{z}{v}} \; e^{-j\pi D_f z(\; f^{P})^2}$, where $\frac{1}{v} = \frac{d\beta}{d\omega}$ and D_f are the group velocity/dispersion coefficient given by $D_f = \frac{1}{2\pi}\frac{d^2\beta}{df^2} = \frac{d}{df}\frac{1}{v}$ (Fig. 2.9).

If the complex amplitude of the incident pulse is $A(0, t) = e^{\frac{-t^2}{\tau_0^2}}$ with $\frac{1}{e}$ half-width at τ_0, then the complex amplitude is $A(z, t) = \sqrt{\frac{j\pi\frac{\tau_0^2}{D_f}}{z - j\frac{\pi}{D_f}\frac{\tau_0^2}{8}}}e^{j\pi\frac{\left(\frac{\left(t - \frac{z}{v_{phase}}\right)^2}{D_f\left(z - j\frac{\pi}{D_f}\frac{\tau_0^2}{8}\right)}\right)}}$. The intensity of this pulse is given by

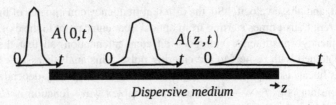

Fig. 2.9 General pulse dispersion

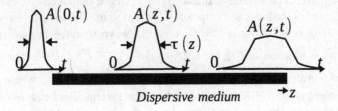

Fig. 2.10 Gaussian pulse dispersion

$$|A(z,t)|^2 = \frac{\tau_0}{\tau(z)} e^{-\left(\frac{t - \frac{z}{v_{phase}}}{\tau(z)}\right)^2}$$

where $\tau(z)$ is the width given by $\tau(z) = \tau_0 \sqrt{1 + \left(\frac{z}{z_0}\right)^2}$, where $z \gg z_0 \tau(z) = \frac{z|D_f|}{\pi \tau_0}$.

This describes the dispersion of a Gaussian pulse (Fig. 2.10).

2.2 Wave Optics: Propagation of Light in Transparent Dielectric Media

Wave optics is the superset of geometrical/ray optics. Geometric/ray optics is the limit of wave optics when the wavelength is infinitesimally short, i.e., so long as light rays travel through media or around objects whose physical dimensions are very large compared to the wavelength of light that geometric/ray optics holds. In the real world, dimensions of most objects are infinitely larger than the wavelength of light that is incident on it, and so wave optics effects are not apparent.

2.2.1 Basics of Monochromatic Light Propagation

Light travels as waves. In free space it travels with speed $c_0 = 3 \times 10^8$ m/s and in a transparent dielectric medium of refractive index $n > 1$, the dielectric constant and refractive index are related as $\epsilon = \frac{c_0}{n}$. The mathematical equation that describes wave propagation is $\nabla^2 u - \frac{\partial^2 u}{c^2 \partial t^2} = 0$ where ∇^2 is the Laplacian operator. This wave equation obeys the *principle of superposition*, i.e., $u\left(\vec{r},\ t\right) = u_1\left(\vec{t},\ t\right) + u_2\left(\vec{r},\ t\right) + \ldots$. At the boundary of two media the wave function changes as a function of the respective refractive indices of the two media. The wave equation is applicable to media whose refractive indices vary linearly and slowly as a function of position, i.e., **locally homogeneous media. The physical meaning of the wave function is not specified, but it allows measurement of real-world parameters as intensity**.

The optical intensity, defined as the light energy per unit area of the incident surface, is directly proportional to the average of the square of the light wave amplitude $U\left(\vec{r},\ t\right) = \left\langle u^2\left(\vec{r},\ t\right)\right\rangle$.. The averaging is done over a time interval that is much larger than the time period of the signal, but much shorter than any other time period of interest (e.g., light pulse duration). The optical power flowing into an area A, as a function of time, is the integral of the intensity over that area $P(t) = \int I\left(\vec{r}.t\right) dA$. By definition, the optical energy is the integral of the optical power over some time interval of measurement.

A monochromatic light wave is represented by a wave function with a periodic sinusoidal time dependency $u\left(\vec{r},t\right) = a\left(\vec{r}\right)\cos\left(\omega\ t + \phi\left(\vec{r}\right)\right)$ where a, ω, ϕ are, respectively, the amplitude, angular frequency, and phase (Fig. 2.11a–c). In complex notation, the above wave function is

$$u\left(\vec{r},\ t\right) = a\left(\vec{r}\right)e^{j\phi\left(\vec{r}\right)}e^{j\omega t}$$

so that $u\left(\vec{r},\ t\right) = U\left(\vec{r},\ t\right) = 0.5\left(U\left(\vec{r}.t\right) + U^{PRIME}\left(\vec{r},t\right)\right)$ and this also satisfies the wave equation defined earlier. The complex wave function can be rewritten as $U\left(\vec{r},\ t\right) = U\left(\vec{r}\right)e^{j\omega t}$ where the time-independent part is the amplitude. This implies that at any location along the wave's propagation, the wave function has a magnitude $U\left(\vec{r}\right)$ and a time dependency. If the complex amplitude is substituted into the wave equation, the Helmholtz's equation is obtained $\left(\nabla^2 + k^2\right) U\left(\vec{r}\right) = 0$ where $k = \frac{\omega}{c}$. The key features of a monochromatic light wave are the following:

- A monochromatic wave is represented by a complex wave function $U\left(\vec{r},\ t\right) = U\left(\vec{r}\right)\ e^{j\omega t}$ *that satisfies the Helmholtz's equation.*

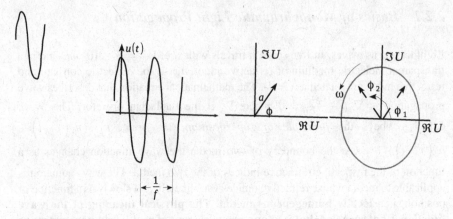

Fig. 2.11 (a) Light wave propagation in time. (b) Complex amplitude and phase angle

- Amplitude and phase of the monochromatic wave function are $\left|U\left(\vec{r}\right)\right|$ and $arg\left(U\left(\vec{r}\right)\right)$, respectively, and the optical intensity is $I\left(\vec{r}\right) = \left|U\left(\vec{r}\right)\right|^2$.

2.2.2 Monochromatic Light Propagation: Plane, Spherical, Paraboloidal, and Paraxial

A monochromatic light propagates in three ways, plane waves, spherical waves, and paraxial waves. A plain monochromatic wave is

$$U\left(\vec{r},t\right) = A\ e^{-j\left(k_x x + k_y y + k_z z\right)}$$

where A is the *complex envelope* and k_x, k_y, k_z are the three components of the wave vector; the magnitude of the wave vector is the wave number. The phase of the plane wave satisfies $arg\left(U\left(\vec{r}\right)\right) = arg(A) - \vec{k}.\vec{r}$ and for this relation to hold, $\vec{k}.\vec{r} = 2\pi q + arg(A)$ where q is an integer. *These relations describe a set of parallel planes whose axes are parallel to the wave vector—plane wave.* The planes are separated by a distance $\lambda = \frac{c}{f}$ called the wavelength. The wave carries infinite power at all locations it is propagating through, and this does not exist in reality (Fig. 2.12).

If the plane wave is propagating through a medium of refractive index n, then the velocity of light in that medium, the wavelength, and the wave vector are related to the corresponding free space values by $c = \frac{c_{VAC}}{n}$, $\lambda = \frac{\lambda_{VAC}}{n}$, $k = \frac{k_{VAC}}{n}$.

Fig. 2.12 Plane wave
propagation

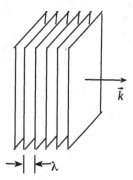

The spherical wave emanates from a point light source, and consists of spheres of ever-increasing radii. The wave function is $U\left(\vec{r}\right) = \frac{A}{r}e^{-jkr}$. The wave intensity is $I\left(\vec{r}\right) = \left(\frac{A}{r}\right)^2$.

A spherical wave originating at $\vec{r}_0$ has a wave equation $U\left(\vec{r}\right) = \frac{A}{|\vec{r}-\vec{r}_0|}e^{-jk|\vec{r}-\vec{r}_0|}$. A variation of the spherical wave is the paraboloidal wave, often called the Fresnel approximation. In Cartesian coordinate system, $r = \sqrt{x^2+y^2+z^2} = z\sqrt{1+\frac{\theta^2}{2}} = z + \frac{x^2+y^2}{2z}$ and then the wave function is transformed to $U\left(\vec{r}\right) = \frac{A}{z}e^{-jkz}\,e^{-j\frac{x^2+y^2}{2z}}$ —an approximation to the spherical wave. A more intuitive approach is the Fresnel approximation.

The underlying physics of the Fresnel approximation is that the spherical wave equation is a plane wave $A\ e^{-jkz}$ modulated by $\frac{e^{-jk(x^2+y^2)}}{2\ z}$ which merely bends the plane waves. Then the wave function is that of just a *paraboloidal wave* that approaches a plane wave when z is very large. The condition for validity of this assumption is that $\frac{N_F\,\theta_{max}^2}{4} \ll 1$ where $\theta_{max} = \frac{a}{z}$ and a is the radius of a sphere centered about the z-axis, and the Fresnel number is $N_F = \frac{a^2}{\lambda z}$ (Fig. 2.13).

A wave is paraxial if the normals to its wave front are paraxial. Such a wave is realized by modulating the complex amplitude of a plane wave so that the envelope varies with position (Fig. 2.14). The complex amplitude then becomes $U\left(\vec{r}\right) = A\left(\vec{r}\right)e^{-jkz}$. To preserve the basic plane wave nature, the variation of $A\left(\vec{r}\right)$ must be very slow compared to its wavelength. The wave function of this wave is $u\left(\vec{r},t\right) = A\left(\vec{r}\right)\cos\left(\omega t - kz + arg\left(A\left(\vec{r}\right)\right)\right)$. As the phase change is very small within one wavelength, the wave fronts bend very slightly, and the normals are paraxial.

Helmholtz's equation for the paraxial waves relies on the condition that $A\left(\vec{r}\right)$ varies very slowly with respect to z, i.e., $\Delta\ A \ll A$ and then

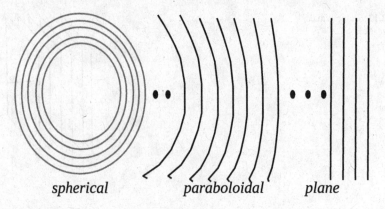

Fig. 2.13 Spherical, paraboloidal (Fresnel approximation), and plane waves

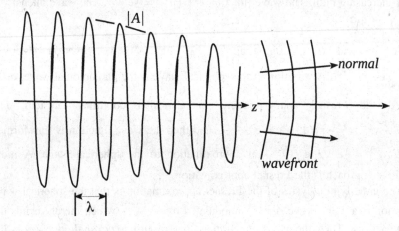

Fig. 2.14 Paraxial waves and wave fronts

$$\Delta\, A = \frac{\partial\, A}{\partial\, z}\Delta\, z = \frac{\partial\, A}{\partial\, z}\lambda$$

and this implies that

$\frac{\partial\, A}{\partial\, z} \ll \frac{A}{\lambda} = \frac{A\, k}{2\pi}\frac{\partial\, A}{\partial\, z} \ll k\, A$. Now combining these results the Helmholtz's equation becomes $\frac{\partial^2 A}{\partial\, x^2} + \frac{\partial^2 A}{\partial\, y^2} - 2j\, k\frac{\partial\, A}{\partial\, z} = 0$.

This is the slowly varying approximation to the Helmholtz's equation or the paraxial Helmholtz's equation.

2.2.3 Common Wave Optic Effects: Reflection, Refraction, Diffraction, and Interference

Two very simple optical real-world effects are reflection and refraction (Fig. 2.15a, b). The boundary condition for reflection is $\vec{k}_1.\vec{r} = \vec{k}_2$, $\vec{r}\forall\vec{r} = \{x, y, 0\}$ where the wave vectors are $\vec{k}_1 = \{k_0 \sin(\theta_{INC}), k_0 \sin(\theta_{INC}).0\}$ and $\vec{k}_2 = \{k_0 \sin(\theta_{REF}), k_0 \sin(\theta_{REF}).0\}$ and θ_{INC}, θ_{REF} are the angles of incidence and reflection, respectively. The boundary conditions for refraction at the boundary of two transparent dielectric media with refractive indices n_1, n_2, respectively, are

$$\vec{k}_1.\vec{r} = \vec{k}_2.\vec{r} = \vec{k}_3.\vec{r} \quad \forall \vec{r} = \{x, y, 0\}$$

where

$$\vec{k}_2 = n_1\{k_0 \sin(\theta_{INC}), n_1 k_0 \sin(\theta_{INC}).0\}$$
$$\vec{k}_2 = n_1\{k_0 \sin(\theta_{REF}), n_1 k_0 \sin(\theta_{REF}).0\}$$
$$\vec{k}_3 = n_2\{k_0 \sin(\theta_{REFR}), n_2 k_0 \sin(\theta_{REFR}).0\}$$

and θ_{INC}, θ_{REF}. θ_{REFR} are, respectively, the angles of incidence, reflection, and refraction.

Another common wave feature is diffraction, **when the amplitude or phase of the incident light wave is modulated periodically, using an optical device called the diffraction grating.** The diffraction grating is a transparent plate with periodically varying thickness on one side, which results in these periodic modulations of the incident wave.

A diffraction grating made of thin transparent material is placed such that a light wave is incident at a small angle θ_{INC} to the normal on the smooth surface. The gratings on the other surface are separated by a distance Λ such that the wave length

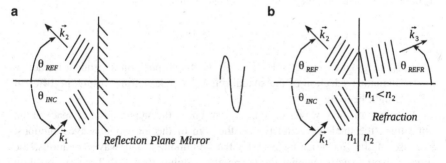

Fig. 2.15 (a) Reflection at a plane mirror. (b) Refraction at the interface of two transparent dielectric media

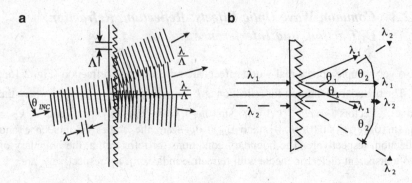

Fig. 2.16 (a) Diffraction with grating and very small angle of incidence. (b) Waves of different wavelengths incident on a diffraction grating

of the incident light satisfies $\lambda < \Lambda$. The angle of diffraction is then $\theta_{DIFF} = \theta_{INC} + k\frac{\lambda}{\Lambda}$ where q is the diffraction order and is given by $q = 0, +/-1, +/-2$. The diffracted waves are separated by $\theta_{SEP} = \frac{\lambda}{\Lambda}$ (Fig. 2.16a).

This set of equations are only valid in the paraxial approximation; that is, the grating period is much greater than the incident light wavelength. In the general analysis, the incident angle is not small, so that $\sin(\theta_{DIFF}) = \sin(\theta_{INC}) + k\frac{\lambda}{\Lambda}$. When two different monochromatic waves (λ_1, λ_2) are incident on a diffraction grating with their propagation vectors parallel to the axis of propagation, the diffracted waves are at different angles*θ_1, θ_2 with respect to the axis of propagation (Fig. 2.16b).

The diffraction grating works by adding a phase shift to the incident wave, which bends as per the incident light at different angles. This phase shift can be controlled by varying the thickness of the grating material along the transverse direction with reference to the optical axis (linearly, quadratically, or periodically). *Identical phase shift can be achieved by having a number of layers of transparent material of very small thickness, each with a slightly different refractive index, fused into one slab.* **This is called the graded index slab and has direct application in graded index optical fibers.** The complex amplitude transmittance of thin planar slab is given by

$$t(v, \ y) = e^{-jn(x,y)k_0 d_0}$$

where d_0, $n(x, y)$ are, respectively, the slab width and position-dependent refractive index. Judicious selection of the spatial refractive index profile results in different types of light bending.

When two or more waves of light are present in the same region of space at the same time, the total wave function is the sum of the individual wave functions. Because of the linear nature of both the Helmholtz's and wave equations, the superposition principle applies to the complex amplitude as well. The superposition principle does not apply to the intensity of the combined wave, and this is because

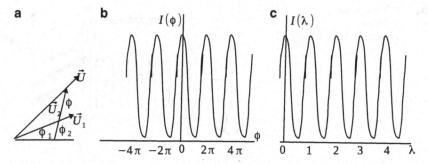

Fig. 2.17 (a) Phasor diagram for interference. (b) Interference intensity with phase angle. (c) Interference as function of path difference in terms of wavelength

the interference of two waves depends on the phase interrelationships of the interfering waves.

When two monochromatic waves interfere, the complex amplitudes add linearly, and the resulting complex amplitude is $U\left(\vec{r}\right) = U_1\left(\vec{r}\right) + U_2\left(\vec{r}\right)$ and the intensity is $U^2 = |U_1 + U_2|^2 = |U_1|^2 + |U_2|^2 + U_1^P U_2 + U_2^P U_1$ where U_1^P, U_2^P are the complex conjugate wave functions. The modulation terms $U_2^P U_1 + U_1^P U_2$ are the interference terms and are the key reason for interference. The term could be positive (constructive interference) or negative (destructive interference). *A phase difference is associated with time delay, so that when the delay distance is an integer multiple of the wavelength, the result is constructive interference. Otherwise, if the delay distance is an integer multiple of half wavelength, the result is destructive interference.* This dependency on phase is exploited to build optical interferometers. Under normal conditions, the random phase relationships (partially coherent light) between the various light waves create random phase interrelationships, which result in intensity patterns that are very difficult to detect. The phasor diagram for two monochromatic light waves is in Fig. 2.17a, the dependency of intensity on phase difference in Fig. 2.17b, and the dependency of intensity on difference distance is in Fig. 2.17c.

When two oblique monochromatic plane waves interfere, an interference pattern is generated, as in Fig. 2.18. One plane wave has one wave vector parallel to the propagation direction (z-axis) while the wave vector of the other is at an angle θ to the z-axis. In this case as well, the modulation terms $U_2^P\ U_1 + U_1^P\ U_2$ generate the interference pattern.

2.2.4 Light Wave Propagation Through Transparent Dielectric Slabs

Optical fibers and slab waveguides transmit light waves from a source to a destination. So this transmission process has to be examined, with a simplification that

Fig 2.18 Interference
between oblique
monochromatic waves

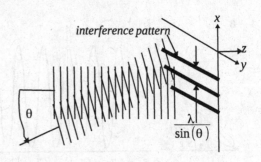

reflection at the entrance is negligible, and a slab is selected as then the analysis is possible in the familiar Cartesian coordinate system.

A monochromatic light wave is incident on a slab of refractive index n and thickness d, with the propagation direction along z. The surfaces at z=0 and z=d are smooth, minimizing reflection, and the complex amplitude U is continuous at the boundaries. The *complex amplitude transmittance* is $\frac{U(x,u,d)}{U(x,y,0)}$. Because of the simplifying assumptions, $\frac{U(x,y,d)}{U(x,y,0)} = e^{-jnk_0 d}$, i.e., thin plate introduces a phase difference $nk_0 d = \frac{2\pi d}{\lambda}$.

If the plane wave is incident at an angle θ with the propagation direction z, the refracted and transmitted waves have wave vectors $\vec{k}_1$, $\vec{k}$ and angles θ_1, θ, respectively, and these angles are related by Snell's law. The complex amplitude inside the slab is proportional to $e^{-j\ \vec{k}_1 \cdot \vec{r}} = e^{-j\ nk_9(z\cos(\theta_1)+x\sin(\theta_1))}$ and the ratio of the complex amplitudes becomes $\frac{U(x,y,d)}{U(x,y,0)} = e^{-j\ nk_0(d\cos(\theta_1)+x\sin(\theta_1))}$.

The effects of reflection at the boundary of the slab and multiple reflections will be examined in detail in later sections.

2.2.5 Polychromatic Light

All of the above analysis is based on the concept of monochromatic wave, which is an idealization. In reality, light waves consist of several frequencies and so are called polychromatic waves. *The general wave function for such a wave is a superposition integral of harmonic functions of different frequencies, amplitudes, and phases:*

$$u\left(\vec{r},t\right) = \int U_f\left(\vec{r}\right) e^{j\omega t} df$$

To ensure that the wave function is real, the complex amplitude must be symmetric, i.e.,

Fig. 2.19 Quasi-monochromatic wave magnitude of Fourier transform of wave function and complex wave function

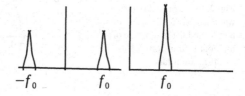

$$\int U_f\left(\vec{r}\right)e^{j\omega t}df = \int U_{-f}\left(\vec{r}\right)e^{-j\omega t}df = \int U_f^P\left(\vec{r}\right)e^{j\omega t}df$$

where the limits of the first integral are $-\infty$, 0, and the second and third 0, ∞. So the wave function, as a function of complex amplitudes, is the integral

$$u\left(\vec{r},\ t\right) = \int\left\{U_f\left(\vec{r}\right)e^{j\omega t} + U_f^P\left(\vec{r}\right)e^{-j\omega t}\right\}df$$

where the limits of integration are 0, ∞. The complex wave function is obtained by first determining its Fourier transform, followed by eliminating all negative frequencies, multiplying the result by 2, and finally determining the inverse Fourier transform. The Fourier coefficients satisfy the wave equation; the complex wave function also satisfies the wave equation. The Fourier transform of a *quasi-monochromatic wave function* has widening of the base, and this is the *phase noise* (Fig. 2.19).

The intensity of a polychromatic wave function is defined as

$$I\left(\vec{r},t\right) = \left\langle u\left(\vec{r},t\right)^2\right\rangle + \left\langle \left(u\left(\vec{r},t\right)^P\right)^2\right\rangle + \left\langle u\left(\vec{r},t\right)u^P\left(\vec{r},t\right)\right\rangle$$

In the special case of the quasi-monochromatic light, the averaging operation eliminates the oscillatory terms in both the complex amplitude and its complex conjugate, at both the positive and negative second harmonic of the fundamental frequency. Other higher order terms are also eliminated by similar techniques, so that the intensity of such light wave is $I\left(\vec{r},t\right) = \left|U\left(\vec{r},t\right)\right|$.

A pulsed plane wave is a polychromatic wave, each of whose components is a plane wave, and its complex amplitude is $U\left(\vec{r},t\right)$ $\int A_f e^{-jkz}e^{j\omega t}df = \int A_f e^{j\omega\left(t-\frac{z}{c}\right)}df$ where the integration limits are 0, ∞ and A_f is the complex envelope. If the phase velocity of the light in the propagation medium is frequency independent, the complex amplitude is $U\left(\vec{r},t\right) = \int A_f e^{j\omega\left(t-\frac{z}{c}\right)}df$. A pulsed plane wave is shown in Fig. 2.20.

The time dependency of the intensity of polychromatic light wave is due to the interference between its plane wave components. If the polychromatic light wave has two frequencies f_1, f_2 then the expression for intensity contains a sinusoidal term with the difference of these two frequencies. This leads to periodic peaks and valleys

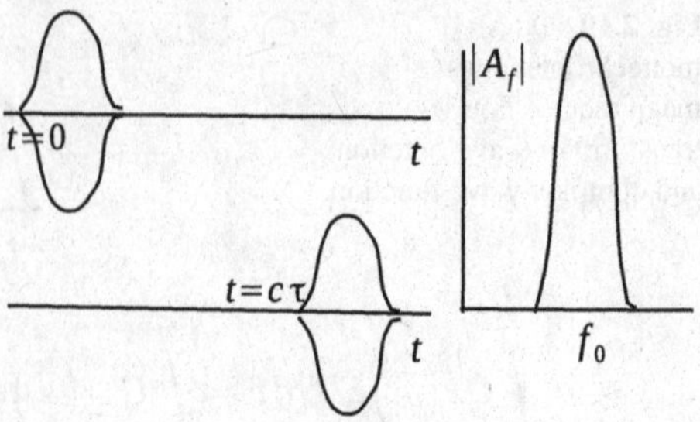

Fig. 2.20 A pulsed plane wave moving at the speed of light c, and the magnitude of the Fourier transform of its complex amplitude

Fig. 2.21 Polychromatic wave consisting of equally spaced (in frequency space) intensity

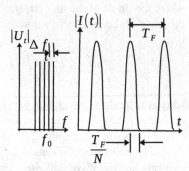

in the light intensity, called "beats" (Fig. 2.10). When a number of monochromatic waves of equal intensities and phases and equally spaced frequencies interfere, very-short-pulse-width, high-intensity pulses are created. If an odd number ($N = 2L - 1$) of these waves interfere, with their frequencies related as $f_q = f_0 + q \Delta f$, where $q =$ -L, ... L, the resulting complex amplitude is the sum of phasors of equal magnitudes with the phase interval $\phi = 2\pi \Delta f t$, i.e., $U(t) = \sqrt{I_0} \sum e^{2j\pi(f_0 + q\Delta)t} - L \leqslant q \leqslant L$ and

the corresponding intensity of this combined wave is $I(t) = |U(t)|^2 =$

$$I_9 \frac{\left(\sin \left(\frac{N\pi t}{T_F} \right) \right)^2}{\left(\sin \left(\frac{\pi t}{T_F} \right) \right)^2}.$$

The intensity pattern is a periodic sequence of pulses, with period $T_F = \frac{1}{\Delta f}$ (Fig. 2.21).

2.2.6 Relation Between Wave and Geometric/Ray Optics

Geometric/ray optics is the limiting case of wave optics when the wavelength is very small, compared to the dimensions of the optical device through which the wave is propagating. The analysis uses monochromatic waves, with free space wavelength λ_0 that is propagating through a locally homogeneous medium. The complex amplitude is $U\left(\vec{r}\right) = a\left(\vec{r}\right)e^{-jk_0 S\left(\vec{r}\right)}$ where $a\left(\vec{r}\right)$, k_0, $S\left(\vec{r}\right)$ are, respectively, the amplitude, free space wave number, and phase. The amplitude varies slowly enough so that it is a constant within any one wavelength. The wave fronts are the surfaces $S\left(\vec{r}\right) = cons \tan t$ and the normals to these surfaces are along the direction of the gradient to these surfaces ∇S. In the immediate space at the end of any position $\vec{r}$ the wave approximates a plane wave, with amplitude $a\left(\vec{r}\right)$, with wave vector $\vec{k}$ parallel to the gradient ∇S and magnitude $k = n\left(\vec{r}\right)k_0$. *Now it is kmown that optical rays are orthogonal to an equilevel surface $S\left(\vec{r}\right) = cons \tan t$ called the eikonal.*

Identifying the local wave vectors (wave front normals) of wave optics with the optical rays of geometric/ray optics, along with constant wave phase planes of wave optics with the eikonal of geometric/ray optics, it is clear how geometric/ ray optics is a special case of wave optics.

Now substituting the above expression for the complex amplitude in the Helmholtz's equation and simplifying, the eikonal equation is obtained as

$$k_0^2\left[n^2 - |\nabla S|^2\right]a\left(\vec{r}\right)_{\grave{\iota}} + \nabla^2 a\left(\vec{r}\right) - jk_0\left[2\nabla S\nabla a\left(\vec{r}\right) + a\left(\vec{r}\right)\nabla^2 S\right] = 0$$

As both terms on the left-hand side of the equation must vanish,

$$n^2 + \left(\frac{\lambda_0}{2\pi}\right)^2 \frac{\nabla^2 a\left(\vec{r}\right)}{a\left(\vec{e}\right)} = \left|\nabla S\left(\vec{r}\right)\right|^2$$

As $a\left(\vec{r}\right)$ varies very slowly with respect to λ_0 it means that $\frac{\lambda_0^2 \nabla^2 a\left(\vec{r}\right)}{a\left(\vec{r}\right)} \ll 1$ and then $\left|\nabla S\left(\vec{r}\right)\right|^2 = \left(n\left(\vec{r}\right)\right)^2$.

Therefore, the scalar function $S\left(\vec{r}\right)$ proportional to the phase in wave optics is the eikonal of geometric/ray optics. In ray optics, $S\left(\vec{r}_B\right) - S\left(\vec{r}_B\right)$ is the optical path difference of two ray positions $\vec{r}_A$, $\vec{r}_B$. Effectively the eikonal equation is the limit of the Helmholtz's equation when the wavelength tends to be zero.

2.2.7 Wave Optics Extension: Beam Optics

Wave optics appears to preclude the idea that propagating waves be confined spatially and traverse only in one direction. However, beam optics enables spatial confinement and propagation in one direction only. ***This is very important to inject light into optical waveguides—optical fibers and transparent dielectric slab waveguides.*** The two main optical wave propagation methods, plane wave and spherical wave, present opposing challenges—while the wave front normals(rays) in case of plane waves are parallel to the propagation direction, the energy carried by the plane wave is spread over the entire wave front, and the normals to a spherical wave diverge from the direction of propagation. The paraxial wave is the only option whose wave front normals make very small angles with the propagation direction. The paraxial wave satisfies the paraxial Helmholtz's wave equation, and thereby satisfies the prerequisites of a beam—**the paraxial wave that satisfies these conditions is the Gaussian beam, as its intensity curve fits the mathematical Gaussian curve (bell shaped) exactly (within applicable tolerances).** Its properties are as follows:

- Wave energy, intensity, and power are confined, mostly to a cylinder about the propagation axis.
- The wave fronts are approximately plane at the beam waist, but become spherical far from the waist.
- The divergence of the wave front normals is minimum as allowed by the wave equation, for a given beam width.

Using the paraxial wave key concepts, the complex amplitude is

$$U\left(\vec{r}\right) = A\left(\vec{r}\right) e^{-jkz} \quad k = \frac{2\pi}{\lambda}$$

where the complex amplitude is approximately constant over a wavelength, so that wave front normals are approximately paraxial. The complex envelope satisfies

$$\left(\frac{\partial^2}{\partial x^2} + \frac{\partial^2}{\partial y^2}\right) A\left(\vec{r}\right) - 2jk\frac{\partial A\left(\vec{r}\right)}{\partial z} = 0,$$

and its simple solution is $A\left(\vec{r}\right) = \frac{A_1}{z} e^{\frac{-x^2+y^2}{2z}}$ A_1 *constant*, and another simple solution is obtained by shifting the reference frame by constant. If the constant is purely imaginary, $A\left(\vec{r}\right) = \frac{A_1}{z+j\,z_0} e^{\frac{-x^2+y^2}{2(z+j\ z_0)}}$ A_1 *constant*, where z_0 is the Rayleigh length. Two real functions, first the measure of beam width R(z) and second the measure of curvature of beam wave front W(z), are introduced to separate the real and imaginary parts of the complex amplitude. Then, $\frac{1}{z+j\ z_0} = \frac{1}{R(z)} - \frac{j\ \lambda}{\pi(W(z))^2}$. Substituting this in the expression for the complex amplitude and rearranging,

$$U\left(\vec{r}\right) = \frac{A_0\,W_0}{W(z)} e^{\frac{-x^2+y^2}{(W(z))^2}}\, e^{\,j\left(\xi(z)-k\,\left(z-\frac{k(x^2+y^2)}{2\ R(z)}\right)\right)} \tag{2.9a}$$

where $W(z) = \sqrt{\frac{\lambda\ z_0}{\pi}}\left(1 + \left(\frac{z}{z_0}\right)^2\right)$, $R(z) = z\left(1 + \left(\frac{z_0}{z}\right)^2\right)$, and $\xi(z) = \arctan\left(\frac{z}{z_0}\right)$.

The properties of the Gaussian beam are deduced easily from the above expressions $I\left(\sqrt{x^2 + y'},z\right) = I_0\left(\frac{\lambda\ z_0}{\pi\ W(z)}\right)^2 e^{\frac{-2(x^2+y^2)}{(W(z))^2}}$ and at each position along the propagation axis z, the intensity profile is Gaussian. The power of a Gaussian beam is simply the integral of the intensity, given by

$$P = \int 2\ \pi\ I\left(\sqrt{x^2 + y^2},\ z\right)\sqrt{x^2 + y^2}d\sqrt{x^2 + y^2}$$
$$= \frac{I_0\lambda^2\,z_0^2}{\pi}\ 0 \leqslant \sqrt{x^2+y^2} \leqslant \infty$$

Along the transverse (to beam propagation direction) the beam intensity is maximum at the beam axis, and falls as $\frac{1}{e^2}$ with the radial distance. Its minimum is in the plane $z = 0$, called the *beam waist*, and the corresponding diameter is the spot size (Fig. 2.22a). $W(z) = \sqrt{\frac{\lambda\ z_0}{\pi}}\left(1 + \left(\frac{z}{z_0}\right)^2\right)$, when $z \gg z_0 W(z) \approx \frac{W_0}{z_0}z = \frac{\lambda}{\pi\ W_0}$, which is also the angular beam divergence. The best beam focus is achieved at $2 = 0$, when its radius is minimum. The distance along the beam axis at which its radius is within $\sqrt{2}\,W_0$ is called the **conformal parameter** or **depth of focus**. The depth of focus is twice the Rayleigh length, $2z_0 = \frac{2\pi\ W_0^2}{\lambda}$.

The phase of the Gaussian beam at any arbitrary position (z, y, z) is $\phi\left(\sqrt{x^2 + y^2},z\right) = k\left(z - \frac{x^2+y^2}{2\ R\ (z)}\right) - \xi(z)$. When phase along the propagation axis alone is considered, the terms involving x and y disappear, and the first term kz is the phase of the plane wave, and the second term represents a phase retardation, which is simply an extra delay compared to the plane or spherical wave. The general case is examined; the term involving x and y is responsible for wave bending when $|z| > 0$ (Fig. 2.22b, c). These key features of the Gaussian beam may be summarized as follows:

- The beam radius is $\sqrt{2}$ times the radius at the beam waist, and the area of the corresponding circle is $2\times$ the area at the beam waist.
- The intensity of the propagation axis is $0.5\times$ peak value.
- The phase of the beam axis is retarded by $\frac{\pi}{4}$ compared to a plane wave.

Understanding the transmission of a Gaussian beam through optical components is very important, since lenses are used to focus Gaussian beams into optical fiber and transparent dielectric slab waveguides. The complex amplitude transmittance of a thin lens with focal length f is $e^{\frac{jk(x^2+y^2)}{2f}}$. A Gaussian beam, while traversing a thin

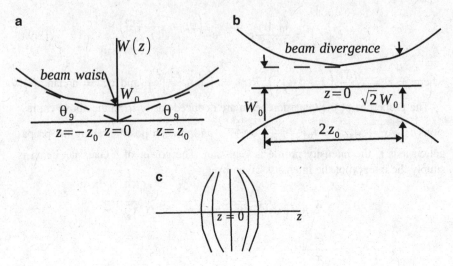

Fig. 2.22 (a) Gaussian beam characteristics. (b) Gaussian beam divergence. (c) Gaussian beam wave fronts

lens, has its complex amplitude multiplied by the phase factor of the lens, with unaltered radius, i.e., a Gaussian beam, with its center at $z = 0$ and waist radius W_0; upon leaving the lens it has its phase altered by $kz + \frac{k(x^2+y^2)}{2}\left(\frac{1}{R(z)} - \frac{1}{f}\right) - \xi(z)$. There are four simplifying parameters for the emerging Gaussian beam:

$$W_0^P(z) = \frac{W(z)}{\sqrt{\left(1 + \left(\frac{\pi(W(z))^2}{\lambda\left(\frac{1}{R(z)} - \frac{1}{f}\right)}\right)^2\right)}}, \quad -z^P = \frac{\left(\frac{1}{R(z)} - \frac{1}{f}\right)}{1 + \left(\frac{\lambda\left(\frac{1}{R(z)} - \frac{1}{f}\right)}{\pi(W(z))^2}\right)^2} \quad r$$

$$= \frac{z_0}{z - f} \quad \text{and} \quad M_r = \left|\frac{z}{z - f}\right|$$

Then magnification $M = \frac{M_r}{\sqrt{1 + r^2}}$, waist radius $W_0^P = M\,W_0$, waist location $(z^f - f) = M^2(z - f)$, depth of focus $2z_9^P = M(2\,z_0)$, and beam divergence $\theta_0^P = \frac{\theta_0}{M}$.

In case $z - f \gg z_0$, i.e., the thin lens is outside the depth of focus of the incident beam, and plane waves turn into spherical waves, resulting in $r \ll 1\,M \approx M_r$, ray optics fails and magnification decreases.

To inject all light energy of an incident Gaussian beam into an optical waveguide, to satisfy the condition, the incident beam dimensions (at the point of entry to the waveguide) must be less than those of the optical waveguide. Therefore, the thin lens must be positioned at the appropriate location with respect to the incident beam, to achieve this goal.

Fig. 2.23 Gaussian beam divergence through thin lens

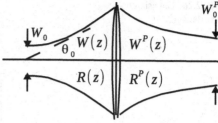

Fig. 2.24 Gaussian beam focusing

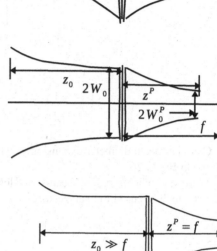

Fig. 2.25 Focusing collimated Gaussian beam

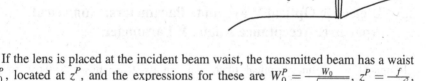

If the lens is placed at the incident beam waist, the transmitted beam has a waist W_0^P, located at z^P, and the expressions for these are $W_0^P = \dfrac{W_0}{\sqrt{1+\left(\frac{z_0}{f}\right)^2}}$, $z^P = \dfrac{f}{1+\frac{f^2}{z_0}}$, where f, z_0 are, respectively, the focal length and location of the incident (before transmission) beam waist (Fig. 2.23). If the depth of focus of the incident beam is much larger than the focal length of the lens, then $W_0^P = \dfrac{f\lambda}{\pi W_9}$ $z^P = f$ (Fig. 2.24).

For applications such as focusing light into an optical fiber, the beam must be focused to the smallest possible size, achieved by using the shortest possible wavelength, thickest incident beam, and shortest focal length lens. The diameter of the lens must collect all the incident light, $D_{lens} \geqslant 2\,W_0$, and the distance of the focused spot is $W_0^P = \dfrac{2f\lambda}{D_{lens}\,\pi}$ (Fig. 2.25).

A number of interesting test cases arise based upon these simple concepts. A Gaussian beam of frequency f can be repeatedly focused by a sequence of lenses if the lens separation is $d \leq 4f$. When a Gaussian beam is focused by a thin lens, the locations of the incident and focused beams are related by

Fig. 2.26 Gaussian pulse
broadening

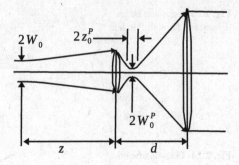

$$\left(\frac{z^P}{f} - 1 \right) = \frac{\left(\frac{z}{f} - 1 \right)}{\left(\frac{z}{f} - 1 \right)^2 + \left(\frac{z_0}{f} \right)^2}.$$

Gaussian beam diameter expansion is achieved by the configuration of lenses as
shown in Fig. 2.26.

There are other types of beams as Hermite-Gaussian and Laguerre-Gaussian
beams which are used to analyze special applications.

2.3 Dielectric Optical Waveguide Parameters: Numerical Aperture, Acceptance Angle, V Parameter

All optical waveguides (optical fiber, slab) in existence now, or those to be
manufactured in future, are made of either high-purity glass or plastic (polymethyl
methacrylate—PMMA) that have solid cores. Each of these solid dielectric core
waveguides has the same physical structure—a core of refractive index n_{core} and an
outer cladding layer of refractive index $n_{cladding}$ satisfying the condition
$n_{core} > n_{cladding}$. Typically, the core and cladding refractive indices are denoted as
n_1, n_2. So, for light to be guided along the waveguide, and then emerge at the other
end after multiple total internal reflections, *the angle of incidence at the core-
cladding boundary* must satisfy the *guided ray condition*:

$$\sin \left(\theta_{inc} \right) > \frac{n_{cladding}}{n_{core}} \text{ or equivalently } \sin \left(\theta_{inc} \right) > \frac{n_2}{n_1} \qquad (2.10)$$

Fig. 2.27 Acceptance and critical angles for optical fiber

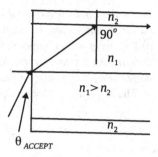

2.3.1   Numerical Aperture, Acceptance Angle, and V Parameter

The *numerical aperture* of an optical fiber is defined as $NA = \sqrt{n_1^2 - n_2^2}$ [1–28]. The analysis for the *acceptance angle* is made on the basis of Fig. 2.27, which illustrates a ray of light incident on an optical fiber at the air-core interface at the optical fiber longitudinal axis, with angle of incidence θ_{accep}. The refractive indices of the core and cladding material are $n_1, n_2\ n_1 > n_2$, respectively. θ_{accep} is special, as the angle of incidence for the same incoming ray of light at the core-cladding boundary is the *critical angle* $\theta_{critical}$, which means that the angle of refraction at the same core-cladding interface is 90°.

From Snell's law applied at the air-core interface,

$$\sin\left(\theta_{accep}\right) = n_1 \sin\left(\theta_{refrac}\right) \tag{2.11}$$

From Snell's law applied at the core-cladding interface,

$$n_1 \sin\left(\theta_{critical}\right) = n_2 \tag{2.12}$$

From the right-angled triangle, $\theta_{refrac} = 90° - \theta_{critical}$,

$$\sin\left(\theta_{refrac}\right) = \sin\left(90° - \theta_{critical}\right), \text{i.e., } \sin\left(\theta_{refrac}\right) = \cos\left(\theta_{critical}\right) \tag{2.13}$$

From Eqs. (2.11) and (2.12) $\sin(\theta_{accep}) = n_1 \cos\left(\theta_{critical}\right)$or

$$\sin\left(\theta_{accep}\right) = \sqrt{n_1^2 - n_2^2}, \text{that is, } \sin\left(\theta_{accep}\right) = \textit{numerical aperture}$$
$$= NA \tag{2.14}$$

Physically, this means that if the angle of incidence of a given ray of light (at the air-core interface) exceeds the acceptance angle, it will not be guided along the optical fiber to emerge at the other end. *Effectively, the acceptance angle forms a cone with its axis coinciding with the longitudinal axis of the optical fiber.*

For a slab dielectric optical waveguide the concept of acceptance angle still holds but acceptance cone is meaningless. For an optical fiber, a low acceptance angle (0.1 ≤ accep tan ce angle ≤ 0.3 in radians) indicates a single-mode fiber, which in turn translates to low attenuation, essential for ultra-long-haul and ultrahigh-bandwidth data transmission.

The V parameter for an optical fiber is defined as

$$V = \frac{2 \pi \, a \sqrt{n_1^2 - n_2^2}}{\lambda_0} \tag{2.15}$$

where a is the fiber core radius, λ_0 the wavelength of the incident light in free space, and n_1, n_2 with $n_1 > n_2$ the refractive indices of core and cladding, respectively. The derivation for Eq. (2.15) involves first formulating the cylindrical coordinate system based on the characteristic equation for the optical fiber, applying the extremely powerful **weak guidance** approximation, and then analyzing the result. The weak guidance approximation is used to circumvent the problem of manipulating Bessel's functions of various kinds and orders and will be examined in detail later.

Another very important parameter is *beam waist*, although it is not directly related to an optical fiber. A number of light sources, e.g., a laser, have Gaussian intensity profile, and the diameter of the beam decreases and increases again. The beam diameter is called the beam waist, and to ensure that maximum light energy is injected into the optical fiber, its input must be placed at the minimum beam waist.

2.4   Optical Fiber Types Graded, Step Index, Single, Multimode Cutoff Frequency

Optical fibers are classified [1–28] in terms of their core refractive index profile, graded or step. The most general case is the graded index, in which case the refractive index profile as a function of core radius is given as

$$n(r) = n_1 \sqrt{1 - 2\left(\frac{r}{a}\right)^p \Delta} \tag{2.16}$$

where n_1 is the refractive index at the core center along the optical fiber longitudinal axis, a is the core radius, p is the grade profile parameter, and $\Delta = \frac{n_1^2 - n_2^2}{2n_1^2}$. $p \to \infty \, n(-r) \to step \, function$, i.e., step index fiber, and when $r = a \, n(a) = n_2$. Therefore the step index fiber is a special case of the general graded index fiber.

Optical fibers are also classified as *single* or *multimode*. The simple test for this is to estimate the V parameter using Equation (2.15) and then check if it is less than **2.405**. If yes, then it is a single-mode fiber, else multimode. The expression for the V parameter can be used to estimate the minimum frequency that can be transmitted by the optical fiber. The corresponding wavelength is easily estimated as

$$\lambda_{cutoff} = \frac{2\pi a\sqrt{n_1^2 - n_2^2}}{2.405} \text{ and then } f_{cutoff} = \frac{c}{\lambda_{cutoff}} \qquad (2.17)$$

where c is the speed of light in m/s.

Refractive index profiles and single multimode properties of optical fibers can be combined to provide single-mode step index, multimode step index, and multimode graded index optical fibers. Multimode fibers suffer from chromatic dispersion [29–41] and are therefore unsuitable for long-haul and high-bandwidth data transmission. Signal energy losses in optical fibers will be examined in detail later.

2.5 Optical Fiber Ray Propagation: Meridional and Skew

Light rays traversing an optical fiber can be *meridional* or *skew* [1–35]. Meridional rays graze any plane through the longitudinal axis of the optical fiber (Fig. 2.28a). The meridional rays intersect the longitudinal axis multiple times and undergo total internal reflection in the same plane, without changing the angle of incidence at the core-cladding interface. This is identical to light propagation in a slab dielectric waveguide, in which case each ray grazes a plane as it traverses a slab waveguide. *Meridional rays propagating through single-mode fibers are the best choice for ultra-long-haul and ultrahigh-bandwidth data transmission, as this combination minimizes signal attenuation.*

A slewed ray grazes any plane **offset** from the optical fiber axis by some distance d less than the optical fiber core radius. The plane is identified by two angles (δ, γ) (Fig. 2.28b). The skewed ray traverses a helical path within a cylindrical shell of inner radius d and outer core radius.

Fig. 2.28 (**a, b**) Meridional and skewed rays in step index optical fiber. Parts of the skew ray on the other offset plane have not been shown

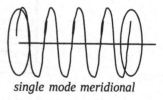

single mode meridional

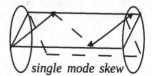

single mode skew

2.5.1 Graded Index Optical Fibers

The refractive index of a graded index optical fiber varies as a function of the radial distance from the axis. Applying the paraxial approximation, light ray propagation is controlled by the paraxial ray equation, with the refractive index given by $(n_{RAD})^2 = n_0^2[1 - (x^2 + y^2)]$. After some manipulation, the following two equations are obtained: $\frac{d^2 x}{d\,z^2} = -\alpha^2\,x$ and $\frac{d^2 y}{d\,z^2} = -\alpha^2\,y$.

These are harmonic oscillator equations with period $T = \frac{2\pi}{\alpha}$, and the initial angles $(\theta_{x0} = \frac{dx}{dz}, \theta_{y0} = \frac{dy}{dz}$ at which light enters the graded index optical fiber) and positions $(x_0,\ y_0)$ determine the amplitudes and phases. The general solutions are $x(z) = \frac{\theta_{x0}}{\alpha}\sin(\alpha\,z)$ and $y(z) = \frac{\theta_{y0}}{\alpha}\sin(\alpha\,z) + y_0\cos(\%alpga\,z)$. The propagation of meridional and skew rays through graded index optical fiber is shown in Fig. 2.29a–c. Figure 2.29c shows the propagation constants and the corresponding confinement volumes for different modes propagating through a graded index multimode optical fiber. *The dashed horizontal lines corresponding to the ends of each vertical dashed line indicate the inner and outer radii of the cylindrical shell through which the wave (corresponding to that mode) propagates.* Each mode is denoted by apir of positive integer numbers – e.g., for the pair 1–4 indicates that the inner and outer radii of the cylindrical shell are r_1, r_2, respectively.

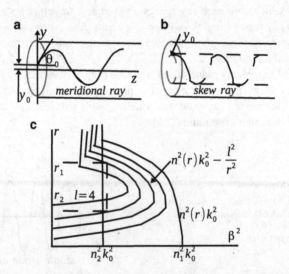

Fig. 2.29 (**a, b**) Meridional and skew rays in a graded index optical fiber. (**c**) Propagation constants and confinement regions in graded index multimode optical fiber

2.6   Optical Fiber Modal Analysis, Transcendental Eigenvalue Equations, Weak Guidance

The analysis of optical fiber propagation modes starts out with the step index optical fiber [1–28]. It can be extended to gradient index fibers as well. Because of its cylindrical symmetry, the electromagnetic waves (light waves) propagating through an optical fiber can be expressed as solutions to Maxwell's classical electrodynamics equations in cylindrical coordinates. **There is a set of allowed special solutions called** *modes*, **each with a unique propagation constant, a unique transverse (to propagation direction) plane electric and magnetic field distribution, and two independent polarization states.**

Each of the electric and magnetic field components must satisfy the Helmholtz's equation $\nabla^2 U + n^2 k_0^2 U = 0$ where $n = n_1$ in the core and $n = n_2$ in the cladding, and the free space wave number is $k_0 = \frac{2\pi}{\lambda_0}$ where λ_0 is the free space wavelength. Recasting operator ∇^2 in the Helmholtz's equation is rewritten as

$$\frac{\partial^2 U}{\partial r^2} + \frac{\partial U}{r \partial r} + \frac{\partial^2 U}{r^2 \partial \phi^2} + \frac{\partial^2 U}{\partial z^2} + n^2 k_0^2 U = 0 \qquad (2.18)$$

Using the separation of variables technique, the target solution is of the form $U(r, \phi, z) = U(r)e^{-jl\phi} e^{-j\beta z}$, which after substitution in (2.18) generates the ordinary differential equation for u(r):

$$\frac{d^2 u}{d r^2} + \frac{du}{rdr} + \left(n^2 k_0^2 - \beta^2 - \frac{l^2}{r^2} \right) u = 0 \qquad (2.19)$$

An electromagnetic wave is guided or bound in the core if the propagation constant is less than the product of the core refractive index and the free space wave number and simultaneously the same propagation constant is greater than the product of the cladding refractive index and free space wave number. That is, $n_2^2 k_0^2 \leq \beta^2 \leq n_1^2 k_0^2$. Now defining $k_T^2 = n_1^1 k_0^2 - \beta^2 \gamma^2 = \beta^2 - n_2^2 k_0^2$, Equation (2.19) is transformed into two separate differential equations, one for the core and the other for the cladding:

$$\frac{d^2 u}{d r^2} + \frac{du}{rdr} + \left(k_T^2 - \frac{l^2}{r^2} \right) u = 0 \text{ (r < a) and } \frac{d^2 u}{d r^2} + \frac{du}{rdr} - \left(\gamma^2 + \frac{l^2}{r^2} \right) u$$

$$= 0 \text{ (r > a)} \qquad (2.20)$$

where $k_T^2, \gamma^2 > 0.0$ and k_T, γ are real. Excluding those solutions for which $u(r) \to \infty$ as $r \to 0$ or $r \to \infty$ the physically viable bounded solutions are the familiar Bessel's functions of the first and second kinds of order $l u(r) \propto J_l(k_T r)$ $r < a$ and where

$$k_T^1 + \gamma^2 = (n_1^2 - n_2^2)k_0^2 \qquad (2.21)$$

The propagating or bounded modes in an optical fiber are obtained by enforcing that the radial solutions (2.21) are continuous at $r = a$. This leads to an eigenvalue equation for β in terms of known parameters as core radius, free space wavelength, and refractive induces of core and cladding. However, deriving the eigenvalue equation involves a number of very complicated mathematical manipulations involving Maxwell's equations. The resulting eigenvalue equation itself is complicated, more so if high-order Bessel's functions of the first and second kinds are used. The easiest and practical solution to this problem is to use the **weak guidance** scheme.

The weak guidance method is based on a very simple but extremely powerful real-world physical fact—the refractive index values of the core and cladding of all real-world optical fibers differ by a few percent. The advantages of this scheme are the following:

- The guided waves are approximately parallel to the fiber axis.
- The longitudinal components of the electric and magnetic fields are very weak compared to the transverse electric and magnetic field components.
- The guided electromagnetic waves are approximately transverse electromagnetic or TEM.
- The linear polarizations in the transverse x and y directions form the orthogonal linearly polarized states of the propagating wave.
- The linearly polarized (l, m) mode is denoted as the $LP_{l\,m}$ mode, and travels with the same propagation constant and spatial distribution.

The vastly simplified eigenvalue equation, in terms of Bessel's functions, takes the form

$$X\frac{J_{l\pm1}(X)}{J_l(X)} = \pm Y\frac{K_{l\pm1}(Y)}{K_l(Y)} \text{ where } X^2 + Y^2$$

$$= V^2 \text{ where V is the V parameter} \qquad (2.22)$$

Although much simplified, the eigenvalue Eq. (2.22) derived by using the weak guidance scheme is still too difficult to be solved, numerically, let alone analytically.

- Both left/right-hand sides of the eigenvalue equation are Bessel's functions. Each of these is an infinite series, so one has to decide on how many terms of each series one would use.
- Both left/right-hand sides of the eigenvalue equation involve polynomial division, a complicated calculation. We might choose to use a mathematical software package as Mathematica, MATLAB, or Mathcad, but for each of these, the user does not know the algorithm used to do the polynomial division and/or any simplifying assumptions that the tool makes.
- The Bessel's functions can be replaced with sine/cosine functions for large values of the argument, but how large is large?

2.7 Weak Guidance, Cutoff Frequency, and Propagating Modes

Careful examination of the eigenvalue Eq. (2.22) shows that whereas the left-hand side is independent of the V parameter, the right-hand side monotonically increases with the V parameter. The minus signs in the eigenvalue equation indicate where the left-hand side intersects abscissa, at the roots (denoted as X_{lm} $m = 1, 2, 3...$) of the equation $J_{l-1}(X) = 0$. **Thus the propagating modes of a given optical fiber are the roots $X_{l\ m} < V_{parameter}$. A mode (l, m) is allowed if $X_{l\ m} < V_{parameter}$ and the cutoff mode occurs when $X_{l\ m} = V_{parameter}$.** The minimum root of $J_{l-1}(X) = 0$ is $X_{0\ 1} = 0$ and the next root is $X_{1\ 1} = 2.405$ for $l = 1$. **If the V parameter is less than 2.405 all modes except for the fundamental, i.e., the $LP_{0\ 1}$, are cut off.**

Optical fibers with large V parameter support a large number of modes, as a large number of roots of $J_l(X) = 0$ can be found in the interval $0 < X < V$. As the Bessel's functions are approximated by sinusoidal functions when $X \gg 1$, the roots can correspondingly be represented as

$$X_{l\ m} = \frac{\pi\left(l + 2m - \frac{1}{2}\right)}{2} \tag{2.23}$$

The corresponding cutoff roots are

$$X_{l\ m\ cutoff} = \frac{\pi(l + 2m)}{2} \tag{2.24}$$

where $l = 0, 1, ...\ m > 1$.

The total number of modes is given by $M = \sum_{l=0}^{l_{ax}} \left(\frac{V}{\pi} - \frac{l}{2}\right)$ which, after some manipulation and allowing for two linear polarizations, becomes

$$M = \frac{4\ V^2}{\pi^2} \tag{2.25}$$

The propagation constants for the guided modes are found by solving the eigenvalue equation, and are expressed as

$$\beta_{l\ m} = \sqrt{n_1^2 k_0^2 - \frac{X_{l\ m}^2}{a^2}} \tag{2.26a}$$

For common case of $V \gg 1$, (2.26a) can be rewritten as

$$\beta_{l\ m} = \sqrt{n_1^2 k_0^2 - \frac{(l+2m)^2 \pi^2}{4\ a^2}}$$ (2.26b)

Using (2.25), (2.26b) can be simplified to

$$\beta_{l\ m} = n_1 k_0 \left(1 - \frac{\Delta(l+2m)^2}{M} \right)$$

where $\Delta = \frac{n_1^2 - n_2^2}{2\ n_2^2}$. The normalized propagation constant is $B = \frac{\frac{\beta^2}{k_0^2} - n_2^2}{n_1^2 - n_2^2}$ and the core and cladding parameters are

$$u = V\sqrt{1-B} \quad v = V\sqrt{B}$$ (2.27a)

The normalized propagation constant expression can be manipulated to express the longitudinal propagation constant in terms of the other quantities B, β, n_1, n_2. Given that B is always positive and its numerical value cannot exceed 1, a simple scheme to compute approximate values for the longitudinal propagation constant consists of making educated approximations of B and computing the corresponding value of the longitudinal propagation constant, with given values of the core and cladding refractive indices. This will be examined in detail in the next chapter devoted to analysis and design examples of optical waveguides.

The corresponding group velocities are given as

$$v_{l\ m\ group} = \frac{c}{\left(1 + \frac{(l+2\ m)^2 \Delta}{M} \right)}$$ (2.27b)

where c is the speed of light in vacuum in m/s $\Delta = \frac{n_1^2 - n_2^2}{2\ n_2^2}$ and M is the total possible number of modes.

A single-mode fiber has two normal modes, called E *(horizontal)* and F *(vertical)* with orthogonal polarizations. An ideal single-mode optical fiber has complete rotational symmetry; the E and F modes are called degenerate, with identical propagation constants $\beta_x = \beta_y = \beta_E = \beta_F$. Therefore, any polarization state injected into such a fiber will be delivered unchanged at the destination. Real-world single-mode optical fibers have microscopic imperfections in them, such as asymmetrical lateral stresses, noncircular cores, and variations in refractive index profiles. These imperfections break the circular symmetry of the ideal fiber and lift the degeneracy of the two modes.

The modes therefore travel with different phase velocities and propagation constants and the difference in their effective refractive indices is called the **optical fiber birefringence** $B = |n_x - n_y|$. The state of polarization in a constant birefringent

Fig. 2.30 (a) Polarization ellipses inside a single-mode optical fiber. (**b, c**) Elliptical core and circular core with bow-tie stress components

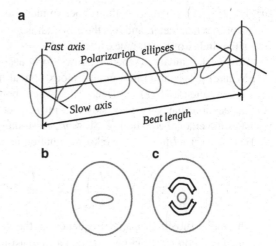

single-mode optical fiber over one beat length implies that the input wave is linearly polarized between the slow and fast axes.

This leads to the concept of polarization-preserving optical fiber. The fiber birefringence is enhanced in single-mode polarization-preserving (polarization-maintaining) fibers, designed to maintain the polarization of the injected wave. Such single-mode optical fibers are called polarization preserving because polarization is preserved in the two normal modes that have different propagation characteristics, preventing them from exchanging energy as they propagate through the fiber. Polarization-preserving fibers are fabricated by including asymmetries into the fiber. Examples include fibers with elliptical cores (which cause waves polarized along the major and minor axes of the ellipse to have different effective refractive indices) and fibers that contain nonsymmetrical stress-producing parts (Fig. 2.30a–c). The bow-tie stress optical fiber works because the bow-tie region is heavily doped with a material that has a different thermal expansion coefficient than the surrounding undoped region. The differential thermal stresses ensure that the core retains its circular shape. This reduces polarization losses. Asymmetries in the elliptical core optical fiber have a similar effect.

2.8 Weak Guidance Gradient Index Profile, Quasi-Plane Waves, Propagation Constants, and Group Velocities for Graded Index Fiber

The weak guidance scheme can be applied to graded index fibers, just like step index fiber. The meridional and skew ray propagation modes are shown in Figs. 2.28 and 2.29. The radial refractive index profile of the graded index fiber is $n(r) =$

$n_1^2\left(1 - 2\left(\frac{r}{a}\right)^p \Delta\right)$ where n_1 is the refractive index at the core center, p is the graded index profile parameter, and a is the core radius. $p \rightarrow \infty$ $n(r) \rightarrow n_1$ which is the core refractive index of a step index fiber.

The weak guidance scheme allows for quasi-plane wave analysis of wave (TEM-like) propagation in a graded index optical fiber. The Helmholtz's equation can be rewritten as $U(r) = a(r)e^{-j k_0 S(r)}$ where both $a(r)$ and $S(r)$ are functions of radial location that vary very slowly with respect to the signal wavelength. $S(r)$ satisfies the eikonal equation $|\nabla S|^2 \approx n^2$. Assuming that the solution is of the form $k_0 S(r) = r u(r) + l \phi + \beta z$, the eikonal equation reduces to

$$\left(k_0 \frac{dS}{dr}\right)^2 + \beta^2 + \frac{l^2}{r^2} = n(r)^2 k_0^2 \tag{2.28}$$

The radial direction spatial frequency of the waver is $k_r = l_0 \frac{dS}{dr}$, and then the assumed solution $U(r)$ can be rewritten as a quasi-plane wave:

$$U(r) = a(r)\left(e^{-\int_0^r k_r\, dr}\right)e^{-jl\phi}\ e^{-j\beta z} \tag{2.29a}$$

which in turn gives

$$k_r^2 = n^2(r)k_0^2 - \beta^2 - \frac{l^2}{r^2} \text{ or } k_r^2 + k_\phi^2 + k_\beta^2 = n(r)^2 k_0^2 \tag{2.29b}$$

That is, the local wave vector $\mathbf{k}$ (of the quasi-plane wave) has three components (k_r, k_ϕ, k_z) in the cylindrical coordinate system. *Most importantly, the condition $k_r > 0$ determines the cylindrical shell region $r_{inner} \leqslant r \leqslant r_{outer}$ of the optical fiber core in which the quasi-plane wave is bound and through which it propagates.* This cylindrical shell region corresponds to the roots of

$$n^2(r)k_0^2 - \beta^2 - \frac{l^2}{r^2} = 0.$$

The modes of the propagating quasi-plane wave in the cylindrical shell (mentioned previously) obey the consistency condition that the propagating quasi-plane wave repeats itself after each helical period in this shell.

- The azimuthal path length corresponding to one complete traversal of the helix must equal a phase shift of a multiple of 2π: $2\pi k_\phi r = 2\pi l$ $l = 1, 2..$
- The radial phase shift for one complete traversal of the helix must equal 2π:

$$2 \int_{r_{inner}}^{r_{outer}} k_r\, dr = 2\pi m \quad m = 0, 1, 2, 3,.$$

The total number of supported modes is the sum of modes for each $l = 0.1.2.3.$ *For each l, the number of modes with propagation constants greater than a given propagation constant value* β *is* $M_l(\beta) = \frac{1}{\pi} \int_{r_{inner}}^{r_{outer}} k_r \, dr$. *Therefore, the total number of* modes whose propagation constants are greater than a specified value is

$$N_\beta = 4 \sum_{l=0}^{l_{max}(\beta)} M_l(\beta) \tag{2.30a}$$

where $l_{max}(\beta)$ is the maximum value of l for a bound mode with propagation mode greater than β. The number 4 accounts for circular polarization of the guided light wave, i.e., the positive and negative polarities of the azimuthal angle ϕ corresponding to the positive and negative trajectories of the same guided or bound light wave, for each (l, m). The above sum can be replaced by an integral, and exploiting the fact that for a graded refractive index core, the mathematical form of the profile is power law, (2.30a) can be evaluated as

$$N_\beta = \left(\frac{p}{2(p+2)} V^2 \right) \left[\frac{1 - \left(\frac{\beta}{n_1 k_0} \right)^2}{2\Delta} \right]^{\frac{p+2}{p}} \tag{2.31}$$

where $V = \frac{2\pi a \sqrt{n_1^2 - n_2^2}}{\lambda_0}$ (V parameter) and $\Delta = \frac{n_1^2 - n_2^2}{2n_1^2}$.

The propagation constant for mode m is $\beta_m = n_1 k_0 \sqrt{\left(1 - 2 \left(\frac{m}{M} \right)^{\frac{p}{p+2}} \right)}$ and the

corresponding expression for group velocity is

$$v_{m \, group} = c \left[1 - \left(\frac{p-2}{p+2} \right) \Delta \left(\frac{m}{M} \right)^{\frac{p}{p+2}} \right] \tag{2.32}$$

In the previous section, the dielectric cylinder waveguide has been analyzed in minute detail, impossible without the weak guidance approximation, and even then a number of simplifying assumptions have to be added, to extract useful and viable expressions for maximum number of modes, propagation constants, etc. The weak guidance approximation is based on the real-world fact that the refractive index of the core material is slightly different(a few percent at most) from the cladding material refractive index. So, effectively there is no alternative to creating an experimental test bench to estimate its performance characteristics to estimate which modes are propagating, and their group velocities.

2.9 Slab Transparent Dielectric Waveguide: Eigenvalue Equation, Even and Odd TE and TM Modes—A

Unlike the cylindrical dielectric waveguide (optical fiber), the dielectric slab waveguide can be analyzed in a straightforward manner in the familiar Cartesian coordinate system (x, y, z). The eigenvalue equations for both even and odd TE and TM modes of propagation can be derived in a straightforward manner, which can be solved numerically using supplied ready-to-use C computer language executables for both the popular Linux and Windows operating systems. There is no need to add any simplifying assumptions.

For TE mode, the amplitude of the electric field vector in the direction of electromagnetic wave field propagation is zero, i.e., $E_z = 0$. For the TE mode, the equation $\frac{d^2 H_z(y)}{d\,y^2} + h^2 (H_z(y))^2 = 0$ must be satisfied in both the core and cladding, with different parameters. In the core, $h^2 = h_{core}^2 = k_y^2 = \omega^2 \epsilon_{core}\mu_0 - \beta^2$ and the general solution to the above differential equation is

$$H_z(y) = H_{odd} \sin (k_y)y + H_{even} \cos (k_y)y \; |y| \leq \frac{d}{2} \qquad (2.33)$$

where even and odd correspond to symmetric and antisymmetric wave functions about $y = 0$. In the cladding, the transverse (orthogonal to the propagation direction) wave function must decay rapidly while penetrating into the cladding material, i.e., $h_{clad}^2 = -\alpha^2 = \omega^2 \epsilon_{clad}\mu_0 - \beta^2$. In both cases, k_y and α are positive. Also, $k_{core} = k_y$. k_{clad} are the wave numbers in the core and cladding, respectively. After applying the boundary conditions to the even and odd wave functions,

$$\alpha = k_y \tan \left(\frac{k_y d}{2} \right) \qquad (2.34a)$$

and

$$\alpha = \sqrt{\omega^2 \mu_0(\epsilon_{core} - \epsilon_{clad}) - k_y^2} \qquad (2.34b)$$

Equating (2.34a), (2.34b) gives the eigenvalue equation for the **even TE modes** of the dielectric slab waveguide:

$$k_y \tan \left(\frac{k_y d}{2} \right) = \sqrt{\omega^2 \mu_0(\epsilon_{core} - \epsilon_{clad}) - k_y^2} \qquad (2.35a)$$

Using similar analysis, the transcendental eigenvalue equation for the **odd TE modes** is

$$-k_y \cot\left(\frac{k_y d}{2}\right) = \sqrt{\omega^2 \mu_0 (\epsilon_{core} - \epsilon_{clad}) - k_y^2} \qquad (2.35b)$$

The eigenvalue equations for the even and odd TM modes are very similar to the eigenvalue equations for the even and odd TE modes, as

$$k_y \tan\left(\frac{k_y d}{2}\right) = \frac{\epsilon_{core}}{\epsilon_{clad}} \sqrt{\omega^2 \mu_0 (\epsilon_{core} - \epsilon_{clad}) - k_y^2} \text{ and } -k_y \cot\left(\frac{k_y d}{2}\right)$$

$$= \frac{\epsilon_{core}}{\epsilon_{clad}} \sqrt{\omega^2 \mu_0 (\epsilon_{core} - \epsilon_{clad}) - k_y^2}$$

Each of these eigenvalue transcendental equations can be solved accurately and fast using a computer. A set of ready-to-use C computer language executables, for both the popular Linux and Windows operating systems, have been supplied with this book, for the interested reader to download and experiment with dielectric optical waveguide design processes and concepts.

2.10   Slab Transparent Dielectric Waveguide: Eigenvalue Equation, Even and Odd TE and TM Modes—B

This scheme is a refinement of the one presented in the previous section. Using the general expression for a vector in a three-dimensional Cartesian coordinate system, $\vec{r}_{x,y,z} = r_x \hat{x} + r_y \hat{y} + r_z \hat{z}$ where $(\hat{x}, \hat{y}, \hat{z})$ are the unit vectors along the three orthogonal axes, the general solution for the TE mode may be written as

$$E(r)_{|x|<d} = \hat{y} \; E_0 [\cos(k_x \, x)] e^{-jk_z z} \text{ or } E(r)_{|x|<d} = \hat{y} \; E_0 [\sin(k_x x)] e^{-jk_z z} \quad (2.36a)$$

$$E(r)_{x>d} = \hat{y} \; E_1 e^{-\alpha_x (x-d)} e^{-jk_z z} \text{ and } E(r)_{x<-d} = \pm\hat{y} \; E_1 e^{\alpha_x (x+d)} e^{-jk_z z} \quad (2.36b)$$

where

$$k_z^2 + k_x^2 = \omega^2 \, \epsilon_1 \mu_0 \text{ and } k_z^2 - \alpha_x^2 = \omega^2 \, \epsilon_2 \mu_0 \qquad (2.36c)$$

Now the boundary conditions need to be applied at the two core cladding interfaces.

- The component of the electric field parallel to this interface (y-axis component) is continuous for all z.
- The component of the magnetic field parallel to the interface (z-axis component) is continuous for all z.

So,

$$E_0[\cos(k_x d) \wedge \sin(k_x d)] = E_1 \tag{2.37a}$$

$$E_0[k_x \cos(k_x d) \wedge -k_x \sin(k_x d)] = -\alpha_x E_1 \tag{2.37b}$$

From (2.37a) and (2.37b)

$$[\tan(k_x d) \wedge -\cot(k_x d)] = \frac{\alpha_x}{k_x} \tag{2.37c}$$

From (2.37c)

$$\alpha_x^2 = \omega^2 \mu_0 (\epsilon_1^2 - \epsilon_2^2) - k_x^2 \tag{2.37d}$$

Combining (2.37c) and (2.37d) the transcendental equations for even and odd modes are

$$\text{Even}\ \tan(k_x d) = \sqrt{\frac{\omega^2 \mu (\epsilon_1^2 - \epsilon_2^2)}{k_x^2} - 1}\ \text{and odd} - \cot(k_x d)$$

$$= \sqrt{\frac{\omega^2 \mu (\epsilon_1^2 - \epsilon_2^2)}{k_x^2} - 1}$$

For the mth TE mode the values of $k_x d$ lie in the range of $(m-1) \times \frac{\pi}{2} \le k_x d \le m\frac{\pi}{2}$ m=1,2,3, ... and to take advantage of this, the above even and odd eigenvalue equations need to be rewritten as even $\tan(k_x d) = \sqrt{\frac{\omega^2 \mu_0 (\epsilon_1 - \epsilon_2) d^2}{(k_x d)^2} - 1}$ and odd $-\cot(k_x d) = \sqrt{\frac{\omega^2 \mu_0 (\epsilon_1 - \epsilon_2) d^2}{(k_x d)^2} - 1}$.

Once the eigenvalue $(k_x d)$ corresponding to a given operating frequency has been determined, the key longitudinal propagation constant (along wave propagation direction) is obtained from

$$k_z = \sqrt{\omega^2 \mu_0\ \epsilon_1 - k_x^2} \tag{2.38}$$

For TM modes, using identical reasoning, the general solution for the TM mode may be written as

$$H(\vec{r})\Big|_{|x|<d} = \hat{y}\ H_0 \cos(k_x x)\ e^{-jk_z z}\ \text{and}\ H(\vec{r})\Big|_{|x|<d}$$

$$= \hat{y}\ H_0 \sin(k_x x)\ e^{-jk_z z} \tag{2.39a}$$

Fig. 2.31 Dielectric slab
waveguide with even
(symmetric) and odd
(antisymmetric propagating
modes)

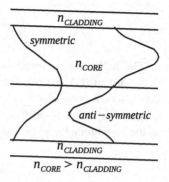

$$H\left(\overrightarrow{r}\right)_{x>d} = \hat{y}\ H_1\ e^{-\alpha_x(x-d)}\ e^{-jk_z z},\ \ H\left(\overrightarrow{r}\right)_{x<-d}$$

$$= -\hat{y}\ H_1\ e^{\alpha_x(x+d)}\ e^{-jk_z z} \tag{2.39b}$$

where $k_z^2 + k_x^2 = \omega^2\ \epsilon_1\ \mu_0$ and $k_z^2 - \alpha_x^2 = \omega^2 \epsilon_2 \mu_0$. After applying the continuity conditions of the core-cladding boundary, one obtains the following eigenvalue transcendental equations for the even and odd TM modes:

Even $\tan\left(k_x d\right) = \frac{\epsilon_1}{\epsilon_2}\sqrt{\frac{\omega^2\mu_0(\epsilon_1-\epsilon_2)d^2}{(k_x d)^2} - 1}$ and odd

$$-\cot\left(k_x d\right) = \frac{\epsilon_1}{\epsilon_2}\sqrt{\frac{\omega^2\mu_0(\epsilon_1 - \epsilon_2)d^2}{(k_x d)^2} - 1}$$

For the mth TM mode, the values of $k_x d$ are in the range of $(m-1)\times \frac{\pi}{2} \leqslant k_x\ d \leqslant m\frac{\pi}{2}$
$m = 1, 2, 3. \ldots$

For both the TE and TM modes, the cutoff frequency is that frequency for which the waves **will not** *be totally internally reflected at the core-cladding boundary.* This means that $\theta_{incident} < \theta_{critical}$ or $\sin\left(\theta_{incident}\right) \leqslant \sin\left(\theta_{critical}\right) = \sqrt{\frac{\epsilon_2}{\epsilon_1}}$ from Snell's law. Or

$$\frac{k_z}{\omega\ \sqrt{\mu_0\ \epsilon_1}} < \sqrt{\frac{\epsilon_2}{\epsilon_1}}\ \text{ or }\ k_z < \omega\ \sqrt{\epsilon_2\,\mu_0}\ \text{ or }\ \omega < \frac{k_x}{\sqrt{\mu_0(\epsilon_1 - \epsilon_2)}} \tag{2.40a}$$

So for example, for the mth node, the smallest value of the transverse propagation constant is $(m-1)\frac{\pi}{2d}$ and substituting this in (2.40a) gives

$$\omega_m = \frac{(m-1)\pi}{2d\sqrt{\mu_0(\epsilon_1 - \epsilon_2)}} \tag{2.40b}$$

The **dispersion curve** for a slab waveguide shows how the longitudinal propagation constant ($\beta \vee k_z$) for different modes varies with the angular frequency ω. It consists of a triangular region whose two boundaries are $k_z = \sqrt{\omega_0 \epsilon_1}$ and $k_z = \sqrt{\omega_0 \epsilon_2}$. Dispersion(to be examined in detail later) occurs in a multimode waveguide as the longitudinal propagation constants of different modes vary.

2.11 Dielectric Slab Waveguide Mode Chart, Mode Numbering, and Waveguide Design

The mode chart for a dielectric waveguide is essential to analyze the propagation of modes through it. The thickness/diameter (d) of the waveguide core is normalized by the wavelength of the incident light. The different modes of propagation are plotted for the propagation angle and different effective refractive indices. For a waveguide with core and cladding refractive indices n_1, n_2, respectively, the critical angle is $\theta_{critical} = \arcsin\left(\frac{n_2}{n_1}\right)$. Most importantly, the angle of incidence at the point of entry into the waveguide (**not** the core-cladding boundary) can vary, starting with the critical angle and up to $90°$, to satisfy the total internal reflection condition. The key information that can be extracted from the mode chart is the following:

- If the core thickness is very small compared to the wavelength of the propagating light the wave travels very close to the critical angle and the effective index approximates the refractive index of the cladding. The wave penetrates deep into the outer layers, because the rays are near the critical angle.
- As the core thickness increases (compared to the wavelength of the incident light) the effective refractive index approximates the refractive index of the core.

The mode chart provides a powerful yet simple scheme to design a dielectric slab waveguide. This scheme exploits the fact that for any given propagation angle there is a set of core thicknesses that will allow rays to propagate. The following equation must be satisfied for the higher modes, where m is a positive integer:

$$\frac{d_m}{\lambda_m} = \frac{d_0}{\lambda_0} \frac{m}{2n_1 \cos(\theta)} \tag{2.41}$$

where d_m, λ_m are the slab thickness and wavelength for the mth mode, d_0, λ_0 are the slab thickness and wavelength for the fundamental, n_1 is the refractive index of the core, and θ is the angle of incidence at the slab entry point (the core-cladding boundary). The ladder operator that enables movement in up and down modes is given by

$$\Delta\left(\frac{d}{\lambda}\right) = \frac{1}{2\,n_1\cos(\theta)} \qquad (2.42a)$$

The total number of supported modes is

$$m_{TOTAL} = 2\frac{d}{\lambda_0}\sqrt{n_1^2 - n_2^2} \qquad (2.42b)$$

As in practice the difference between the refractive indices of the core and cladding is very small, the TE, TM, and hybrid modes can be expressed in terms of the linearly polarized modes.

The TE and TM modes are numbered using the number of zeros in their electric field pattern across the waveguide (transverse direction). *In this scheme TE_0 mode has a continuous distribution with a single maximum and no zeros*. Correspondingly, a $TE_{0\,0}$ mode (for a two-dimensional waveguide structure) and its electric field distribution have a single spot (maxima). Due to invariable asymmetries in the waveguide construction, $TE_{2\,1}$ has an electric field pattern with two zeros in one direction and a single zero in the perpendicular direction. The linearly polarized modes, used exclusively in the analysis of optical fibers (cylindrical cross-section dielectric waveguide, as opposed to slab dielectric waveguide), are numbered differently. *The linearly polarized modes are described by LP_{lm} where m is the number of maxima along the radius and l is the maxima along the circumference, respectively, of the fiber.*

2.12   Energy Loss Mechanisms in Optical Waveguides Absorption and Attenuation

There are two main signal energy loss mechanisms [34–42] in optical waveguides— absorption and attenuation. All real-world optical waveguide materials are nonideal (anisotropic, nonlinear, inhomogeneous, etc.), resulting in signal energy *loss. The consequence of signal energy loss is noticeable for optical fibers as compared to optical slab waveguides as the size of optical fibers used for common applications is huge compared to that of an optical slab waveguide.* So, on a long-haul optical fiber link, amplifiers need to be placed at predetermined intervals to maintain uniform signal energy levels between transmitter and receiver. That is, the signal power at any arbitrary position along the fiber axis is $P(z) = P(0)e^{-\alpha_p z}$ where α_p is the attenuation constant in units of $\frac{dB}{km}$. Another common unit is dBm where $dBm = 10\log(P\ milliWatts)$.

Attenuation in an optical fiber is a result of signal energy loss due to absorption and scattering. Absorption occurs when incident light is absorbed and converted to heat by molecules in the optical waveguide core material. Primary absorbers are residual OH+ and dopants used to modify the refractive index of the glass. This absorption occurs at discrete wavelengths, determined by the elements absorbing the light. The OH+ absorption is predominant, and occurs most strongly around 1000 nm, 1400 nm, and above 1600 nm.

The chief cause of attenuation is scattering, and it occurs when light collides with individual atoms in the glass. Glass, a supercooled liquid, has no regular crystalline structure, making it anisotropic. Moreover, no matter how tightly controlled the manufacturing environment is, some minute quantities of impurities are bound to be added. Light scattered at angles outside the numerical aperture of the fiber will be absorbed into the cladding or transmitted back towards the source. Scattering is also a function of wavelength, proportional to the inverse fourth power of the wavelength of the light. Therefore if the wavelength is doubled, the scattering loss is reduced 16 times. Therefore, for long-distance transmission, it is advantageous to use the longest practical wavelength for minimal attenuation and maximum distance between repeaters. Together, absorption and scattering produce the attenuation curve for a typical glass optical fiber. Fiber-optic systems transmit in the "windows" created between the absorption bands at 850 nm, 1300 nm, and 1550 nm, where physics also allows one to fabricate lasers and detectors easily. Plastic fiber has a more limited wavelength band, which limits practical use to 660 nm LED sources. Therefore, the attenuation rate versus wavelength provided by optical fiber manu-facturers is a superposition of the absorption and attenuation curves.

2.12.1 Electromagnetic Wave Attenuation in Dielectric Waveguides

The dielectric constant of a dielectric material has an imaginary component, whose effects become noticeable at resonant frequencies. Using classical electrodynamics arguments, the equation for a plane wave traversing a nonmagnetic linear dielectric material, in the z direction, is $E = E_0 \ e^{j(k\ z\ 0\ \omega\ t)}\ k = \omega\sqrt{\epsilon(\omega)\mu_0}$, and taking into account the frequency dependency of the dielectric constant, the wave number has a real and imaginary part as

$$k = \beta + j\frac{\alpha}{2} \tag{2.43a}$$

Then the propagating plane wave becomes

$$E = E_0 \ e^{-\frac{\alpha}{2}z}e^{\ j(\beta\ z-\omega\ t)} \tag{2.43b}$$

The intensity of this wave, which is the main measurable quantity, is

$$I = \sqrt{\frac{\epsilon}{4}\frac{}{\mu_0}}E_0^2\,e^{-\alpha z} \tag{2.44}$$

Now, equating the above two expressions for the wave number k, one obtains

$$\beta^2 - \frac{\alpha^2}{4} + ja\beta = (\mathfrak{R}(\epsilon) + \mathfrak{I}(\epsilon))\mu_0\,\omega^2 \tag{2.45a}$$

Then,

$$\alpha = \omega\sqrt{2\mu_0}\sqrt{(\epsilon) - \mathfrak{R}(\epsilon)} \ \text{ and } \ \beta = \omega\sqrt{2\mu_0}\sqrt{(\epsilon) + \mathfrak{R}(\epsilon)} \tag{2.45b}$$

The powerful conclusions that can be drawn from Eqs. (2.45a) and (2.45b) are

$\alpha = 2\mathfrak{I}(k)$ Attenuation
$\beta = \mathfrak{R}(k)$ Refraction—higher values mean the wave bounds more at interfaces
$\mathfrak{R}(\epsilon)$ Energy storage in dielectric medium
$\mathfrak{I}(\epsilon)$ Energy loss in dielectric medium

2.12.2 Electromagnetic Wave Energy Loss in Optical Waveguides Scattering

Scattering occurs in a dielectric optical waveguide when light incident on atomic size impurities excites the impurity particle to higher energy state, and the particle in turn reradiates the absorbed energy as it returns to its ground energy state—the simplest and yet powerful quantum mechanical explanation. A special case is surface scattering, which occurs for high-order modes only. For a dielectric optical waveguide of length l and thickness t the number of reflections from each surface is $N_R = \frac{l}{2t\cot(\theta_m)}$ where θ_m is the angle that the electromagnetic wave ray makes with the reflecting surface. The signal intensity is $I(z) = I_0\,e^{-\alpha z}$ and the loss is $L\left(\frac{dB}{cm}\right) = 4.3\,\alpha$.

Scattering from atomic and molecular impurity particles is of two types, depending on the approximate impurity size. If the impurity particle size is less than 10th of the wavelength of the electromagnetic wave, it is Rayleigh scattering, or else Mie scattering. While Rayleigh scattering is uniform, Mie scattering has a front lobe, as in dipole radiation. Rayleigh scattering intensity is: $I = I_0\frac{8\pi^4 N\alpha^2}{\lambda^4 R^2} \times \left(1 + (\cos(\theta))^2\right)$ where α is the polarizability, N is the number of scatterers, and R is the distance from the observer. The Rayleigh scattering field loss coefficient is $\frac{4\pi^3}{3\lambda^4}n_0^9\rho^2\beta_T k_B T$ where n_0 is the average refractive index, ρ the photoelastic coefficient, k_B Boltzmann's constant, and T the absolute melting temperature of glass in Kelvin.

2.12.3 Interband Absorption in Dielectric Optical Waveguide Material

Interband absorption occurs when an electron or hole in a fixed, nonconducting energy band absorbs energy from a photon and is excited to a conducting band. This is in contrast to interband absorption, when the electron or hole is already in an excited energy band.

2.13 Optical Fiber Lasers: Countering Optical Waveguide Absorption and Attenuation

Optical waveguide signal energy loss due to absorption and attenuation is insignificant for slab optical waveguide, as their length is very small (e.g., 5 mm). However, for a long-haul ultrahigh-bandwidth optical fiber (e.g., 50 km), the absorption and attenuation of signals reduce the output signal power levels significantly. To counter this energy loss, the optical fiber laser, in the form of a rare earth (erbium, praseodymium, etc.)-doped optical fiber, is used. There are several advantages to this scheme.

- Light amplification is in situ, thereby eliminating conversion to intermediate electronic signals, followed by their amplification, and final reconversion to optical signals.
- Amplification is independent of the data rate, a consequence of the in situ light amplification.
- Efficient energy transfer ensures that maximum pump laser light is injected into the optical fiber.
- The gain is mostly flat, so that amplifier stages can be connected in series.

The key to boosting signal energy levels in an optical fiber is the fiber-optic laser, based on the three-level model of the laser. A typical long-haul optical fiber core is made from germanium or silicon dioxide, and is doped with rare earth atoms as erbium or praseodymium. As glass is a supercooled liquid (glass molecules do not have any regular, periodic crystalline structure), it results in material transparency, high optical damage threshold, etc. Alkaline and rare earth alkaline oxides cannot form glass by themselves, but can form glassy structure with glass, and are therefore called network modifiers. Pure glass can be processed only at very high temperatures, but addition of network modifiers reduces the melting temperature of glass considerably.

Very simple quantum mechanics underlies the three-level (ground, intermediate, and high) theoretical laser model (Fig. 2.32a–c). The pump laser provides the pump radiation at two wavelengths (980 nm and 1480 nm) to boost the energies of the erbium or praseodymium impurities to the upper and intermediate energy levels. The lifetime of dopants at the high energy level is very short, e.g., 10 ns, and they return

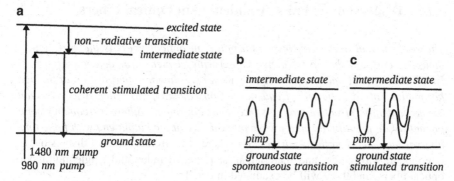

Fig. 2.32 (a) General three-energy-level model of optical amplifier. (b, c) Spontaneous emission transition and coherent in-phase, lockstep stimulated emission

to the intermediate energy level, ***non-radiatively (no emission of any radiation).*** **Then the dopants that have returned to the intermediate level (from the high-energy state) and the dopants initially boosted to the intermediate energy state return to the ground energy state *in phase*. This *in-phase* transition from the intermediate to the ground energy state is accompanied by emission of energy—** *stimulated emission*—**laser light.** The laser action is shown in Fig 2.6a–c. Available literature on laser physics provides detailed descriptions of rate equations, coefficients of spontaneous and stimulated emission, etc.

The equations that govern the rate of change of laser photons in the cavity are simple $\frac{dq}{dt} = q(Gn - b)$ $\frac{dn}{dt} = p - q(Gn + f)$ where q is the total number of photons in the cavity, n the total number of atoms in the excited state, G the gain coefficient, f the decay rate due to stimulated emission, b the decay rate due to transmission and scattering, and p the pump strength. The term Gnq describes the stimulated emission and the more excited atoms in the cavity are, the more photons will be produced, leading to increased stimulated emission. The term bq quantifies the loss mechanism, as energy loss from the cavity occurs only when photons leave the cavity, by scattering or transmission. The second equation describes the time rate of change of photons, where the only gain term is the external pumping rate that increases the number of excited photons inside the cavity. Spontaneous emission is a loss term since it is a random process that decreases the number of photons in the cavity. The significance of the Dnq term can be understood by setting all other terms in the above two equations to zero. Then the above two equations would be reduced to $\frac{dq}{dt} = Gnq$ $\frac{dn}{dt} = -Gnq$. This set of equations imply that a decrease in the number of excited atoms directly relates to an increase in the number of laser photons (adding photons is the only way for the excited atoms to change in this case). This is exactly the correct physical procedure that is needed to describe stimulated emission.

2.14 Dispersion or Pulse Broadening in Optical Fibers

The key cause of dispersion of optical pulses in optical fibers is the variation of dielectric constant (hence refractive index) with frequency (or equivalently wavelength). An optical pulse (e.g., a square wave) is a linear superposition of the odd harmonics of a selected frequency. Consequently, as dielectric constant (or equivalently refractive index) is frequency dependent, different frequency components of an input light square wave pulse arrive at the output end with finite time interval gaps between them—dispersion. There are different types of dispersion, as intermodal, waveguide, and material, in step/graded index and single/multimode optical fibers, and these will be examined in detail.

2.14.1 Intermodal Dispersion or Group Delay

This occurs only in step/graded refractive index core multimode fibers. Each of these modes, because of their slightly different longitudinal propagation constants, traverses along different paths as they traverse the length of the optical fiber. A single-mode optical fiber carries one mode, and so by definition there is no modal dispersion.

Consider a step index multimode optical fiber with two modes of an incident light beam, one propagating parallel to the fiber axis, and the other at angle (less than the acceptance angle) to the fiber axis. The second mode will undergo a very large number of total internal reflections at the core-cladding boundary, before exiting the fiber. So, there will be a very small but finite time interval between the two modes exiting the fiber—intermodal dispersion. The time delay is given as

$$\Delta\, \tau_{INTERMODAL} = \frac{L(n_1 - n_2)}{c}\left(1 - \frac{\pi}{V}\right)$$

where c is the speed of light in vacuum, L is the length of the optical fiber in km, and V is the V parameter.

The analysis for a graded index multimode optical fiber is very complicated, and intermodal dispersion has to be estimated experimentally. In general, the dispersion of the mth order mode is

$$\tau_{MODAL} = \frac{L\, N_{g1}}{c}\left(1 + \frac{g-2-\epsilon}{g+2}\Delta\left(\frac{m}{N}\right)^{\frac{g}{g+2}} + high\ order\ terms\ of\ \Delta\right)$$

where g is graded index profile parameter, c is the speed of light in vacuum, $N_{g1} = n_1 - \lambda\,\frac{dn_1}{d\lambda}$, $N = (akn_1)^2\Delta\left(\frac{g}{g+2}\right)$, $\epsilon = \frac{-2n_1\lambda}{N_{g1}\Delta}\left(\frac{d\Delta}{d\lambda}\right)$, $\Delta = \frac{n_1^2-n_2^2}{2\ n_1^2}$, n_1 is the core center, and n_2 is the cladding refractive index. So, direct experimental estimation of the variation of the core refractive index with wavelength must be made to calculate the modal dispersion. If the core center refractive index profile is set to an optimum value, then $g = g_{opt} = 2 + \epsilon$ and then the expression for modal dispersion can be rewritten as $\Delta \tau = \frac{n_1 \, L \Delta \left(g - g_{opt} \right)}{c(g+2)}$ $g \neq g_{opt}$ and $\Delta \tau = \frac{n_1 \, L \Delta^2}{2c}$ $g = g_{opt}$.

2.14.2 Material Dispersion

As mentioned earlier, dispersion in optical fibers is mainly because the dielectric constant or equivalently the refractive index depends on the wavelength. The group velocity of a pulse of light traversing a homogeneous medium waveguide is $\frac{1}{v_g} = \frac{n(\lambda) - \lambda \frac{dn}{d\lambda}}{c}$ where the expression in the numerator of the right-hand side is called the *group refractive index*, $N(\lambda) = n(\lambda) - \lambda \frac{dn}{d\lambda}$. If the incident light has a spectral width $\Delta \lambda$ then the group delay of each wavelength component is $\Delta \tau = \frac{-L\lambda^2}{c\lambda} \, \Delta \lambda \, \frac{d^2 n}{d\lambda^2}$, and so the material dispersion is specified as $D_m = \frac{-\lambda}{c} \frac{d^2 n}{d\lambda^2} 10^9$ $\frac{ps}{km \, nm}$. The plot of $\frac{d n^2}{d\lambda^2}$ vs. wavelength has a zero crossing point, called zero material dispersion wavelength (ZMCW).

2.14.3 Waveguide Dispersion

The condition for single-mode operation of an optical fiber is that the V parameter is less than 2.405. Waveguide dispersion is relevant only for single-mode fibers, and arises because the propagation constant is a function of the ratio of the fiber radius to the wavelength (since the V parameter is a function of $\frac{a}{\lambda}$), especially for single-mode optical fibers with radius >1.27 micron. The normalized propagation constant of a mode is $b = \frac{\frac{\beta^2}{k^2} - n_2^2}{n_1^2 - n_2^2}$ and for a step index fiber, the normalized propagation constant depends only on the V parameter, and combined with the weak guidance approximation, $\beta = \frac{\omega}{c} \left(n_2 - (n_1 - n_2)b(V) \right)$ and the corresponding group velocity is $\frac{1}{v_g} = \frac{1}{c} \left(n_2 - (n_1 - n_2)b(V) \right) + \frac{\omega}{c} (n_1 - n_2) \frac{db}{dV} \frac{dV}{d\omega}$. Using the expression for the optical fiber V parameter and the weak guidance approximation, the waveguide dispersion in a single-mode optical fiber for a light source of spectral width $\Delta \lambda$ is $\Delta \tau = \frac{-L}{c} n_2 \Delta \left(\frac{\Delta \lambda}{\lambda} \right) \left(V \frac{d^2 (bV)}{dV^2} \right)$. This expression can be evaluated analytically only for the lowest order modes; everything else requires experimental evaluation. The total dispersion for a single-mode fiber is a sum of the material and waveguide dispersions.

The combined effect of material and waveguide dispersion is called *chromatic dispersion*. Chromatic dispersion arises from the frequency dependency of both the optical fiber and waveguide refractive indices; while smaller than the material dispersion, waveguide dispersion shifts the wavelength for the minimum chromatic dispersion. This combined dispersion is very important for single-mode fibers, so that graded index fibers are needed to curb this problem. That is, the core graded refractive index profile is selected so that the wavelength at which the material dispersion compensates the waveguide dispersion is shifted to the wavelength at which the fiber is to be used—the **dispersion-shifted optical fiber**. This is achieved by using a linearly tapered core refractive profile with a reduced core radius. The drawback is that creating an optical fiber with graded index profile requires adding dopants in precisely measured quantities, increasing processing|production cost.

To understand why pulse broadening occurs in multimode fibers due to material dispersion, expressions for the propagation constant and group velocity may be considered, given that both core and cladding refractive indices are frequency dependent. The propagation through graded indes optical fiber with frequency dependent core and cladding refractive indices requires that; the group velocity for the qth mode is $v_q = \frac{d\omega}{d\beta_q} = \frac{c_0}{N_1}\left[1 - \frac{p-2}{p+2}\Delta\left(\frac{q}{M}\right)^{\frac{p}{p+2}}\right]$, where the group index of fiber is $N_1 = \frac{d}{d\omega}n_1 \ \omega = n_1 - \frac{dn_1}{d\lambda_0}\lambda_0$.

In some cases with very-high-intensity light, the refractive indices become intensity dependent, resulting in very interesting nonlinear effects as optical solutions.

2.14.4 Zero and Nonzero Dispersion-Shifted Fiber

The wavelength for zero dispersion is 1300 nm, and the minimum optical fiber loss is around 1449 nm. So, if a single-mode optical fiber is designed to operate beyond 1550 nm, it would support ultrahigh bandwidths and will have very low loss and dispersion over very long distances. This is called the dispersion-shifted optical fiber. However, this type of fiber is useless for wavelength division multiplexing (WDM). When multiple optical channels pass through a zero dispersion-shifted fiber at wavelengths where dispersion is very close to zero, they suffer from a type of cross talk called *four-wave mixing*. The degradation is severe for dense WDM systems.

To tackle this issue, the zero dispersion wavelength is shifted to beyond the wavelength band used by erbium-doped optical amplifiers. In this case, the dispersion is not zero but a very small controlled value.

2.15 Coupled Modes Theory and Integrated Optical Devices

It has been seen that while traversing a dielectric waveguide, an electromagnetic wave (light) penetrate the cladding material, where it decays exponentially in the transverse or radial direction. That is why the cladding of optical fibers is made thick. But as all dark clouds have a silver lining, this effect can be exploited to design optical multiplexers, switches, etc. [42–47]. The coupling between modes can occur between modes propagating in the same waveguide (in the same or opposite directions) or in between two adjacent, *very* closely located waveguides. Each of these cases can be analyzed using perturbation theory. The analysis assumes the weak guidance approximation.

2.15.1 Propagation in Unperturbed Medium and Added Perturbation

Each mode propagating along the longitudinal z-axis can be described by $\vec{E}(x,y)e^{j(\omega t - \beta z)}$ where ω, β are the angular frequency and longitudinal propagation constant, respectively. These are solutions to the equation $[\Delta_T + \omega^2 \eta \, \epsilon_0]\vec{E}(x,y) = 0$. Each solution is orthogonal and so any two modes k and l satisfy $\langle k|l \rangle = \frac{2\omega\mu}{|\beta|}\delta_{kl}$ where $\delta_{k\,l} = 0\, k \neq l\, \delta_{k\,l} = 1\, k = l$ is the Dirac delta function. The propagating wave is a linear superposition of these modes.

When a perturbation is introduced, the propagation equation is changed with the addition of a source term on the right-hand side, as $[\Delta + \omega^2 \, \eta \, \epsilon_0(x, \, y)]\vec{E}(x, \, y) = -\omega^2 \mu \delta \epsilon(x,y,z)\vec{E}(x, \, y)$. In the first approximation, the solution to the perturbed equation is the linear superposition of the unperturbed waves, with an additional z coordinate-dependent coefficient as $\vec{E}(x,y,z,t) = \sum_l A_l(z)\vec{E}(x, \, y)e^{j(\omega t - \beta z)}$. To solve this perturbed equation, the perturbed solution is applied to the perturbed differential equation, and the resulting expression, after some manipulation, gives

$$\sum_l \vec{E}(x, \, y)e^{-j\beta z\left[\frac{d^2 A_l(z)}{d\,z^2} - 2j\beta\frac{d\,A_l(z)}{dz}\right]} = -\mu\,\omega^2 \, \delta \, \epsilon(x,y,z)\sum_l E_l(x,y)e^{-jbeta_l z}\, A_l(z)$$

(2.46)

This is a very difficult expression to solve in a general case. So special cases will be examined in detail that make the expression more tractable to be manipulated. From the orthogonality condition of the propagating modes, it follows that

$$\langle k|k \rangle \left[\frac{d^2 A_k(z)}{d\, z^2} - 2j\beta \frac{d\, A_k(z)}{dz} \right] = -\mu\, \omega^2 \sum_l \langle k|\delta\epsilon(x,y,z)|l \rangle A_l(z) e^{-j(\beta_i - \beta_k)z} \quad (2.47)$$

The coupling between modes k and l is controlled by the coupling coefficient given by

$$C_{kl} = \langle k|\delta\epsilon(x,y,z)|l \rangle$$

2.15.2 Slowly Varying Envelope Approximation

Weak coupling between the modes k and l can be analyzed using the slowly varying envelope approximation. In this case, the envelope of the amplitudes of the coupled modes $A_k(z)$ varies as $\frac{1}{\beta_k}$. Applying this condition means that the second-order differntial in (2.15.1.2) can be ignored and then $\left| \frac{d^2 A_k(z)}{d\, z^2} \right| \ll \left| \frac{\beta_k d\, A_k(z)}{d\, z} \right| \ll \beta_k^2$. This means that for any mode k the second-order differential in (2.15.1.2) can be ignored, and the expression for coupling constant remains as before.

2.15.3 Resonant Coupling

Let only two modes k and l couple—this could be the case when only these two modes propagate, or the coupling constant for all other mode pairs is zero. The difference in $A_k(z)$ between z=0 and z=L (the two waveguide extremities) is very small, so that

$$A_L(L) - A_k(0) \simeq -j \frac{|\beta_k|}{\beta_k} C_k\,_l A_l(0) \int_{L \gg \frac{1}{\Delta\beta}} e^{-j\Delta\beta z dz} \quad (2.48)$$

where the value of the integral is $\frac{\sin\left(\frac{\pi\Delta\beta L}{2}\right)}{\frac{\pi\Delta\beta L}{2}}$. This is the sine function, and has a maximum at $\Delta\beta = 0$. *This is the resonant coupling condition; that is, physically, there is maximum coupling between two modes if and only if their longitudinal propagation constants are the same.* There are two types of perturbations. In the first case the perturbation is invariant along the longitudinal propagation axis. The more interesting and common case is when the perturbation varies along the longitudinal propagation axis. The coupling constant is the sum of the individual coupling constants as $C_k\,_l = \sum_m C_k^{(m)}\,_l e^{\frac{-2j\pi m z}{\Pi}}$. This implies that for two resonantly coupled propagating modes, k and k', the longitudinal propagation constants are related as $\beta_k = \beta_k + \frac{2\pi m}{\Pi}$ where Π is the coupling period.

2.15.4 Coupled Mode Theory Application: Optical Waveguide Coupling

Light propagates through waveguides (optical fibers, transparent slab dielectric waveguides) in modes, whose electrical field complex amplitude is a linear superposition of the propagating modes expressed as $E(y, z) = \sum a_m u_m(y) e^{-j\beta_m z}$, where a_m, β_m, u_m are, respectively, the amplitude, propagation constant, and real transverse distribution, for mode m. The amplitudes of the different modes depend on the energy used to excite the waveguide. A mode is excited in the waveguide, if the source has a distribution that matches this mode. A source with arbitrary distribution excites modes of different strengths in the waveguide. The fraction of the power transferred by the source to the mode m in the waveguide depends on the degree of similarity of s(y) and $u_m(y)$, and is expressed as $s(y) = \sum a_m u_m(y)$. The amplitude of the excited mode l is $a_l = \int a_l u_l(y) d y$ $\infty \leqslant y \leqslant \infty$. The coefficient represents the correlation between the source and mode distributions.

Light is coupled(injected) into an optical waveguide using various techniques (Fig. 2.32a–d). *A mode is excited in the waveguide if the transverse distribution of the incident light matches that of the waveguide mode.* The polarization of the incident light must match that of the incident light. The extremely small dimensions of a typical optical waveguide make the coupling difficult and thus inefficient. For a multimode optical fiber, the guided waves are confined by an angle $\overline{\theta_{CRITICAL}} =$ arccos $\left(\frac{n_{CLADDING}}{n_{CORE}} \right)$, because the acceptance angle θ_{ACCEPT} satisfies the following relation with the numerical aperture of the waveguide $\theta_{ACCEPT} = n_1 \sin(\theta_{CRITICAL}) = \sqrt{n_1^2 - n_2^2} = numerical\ aperture$. The incident angle greater than θ_{ACCEPT} will reduce the coupling efficiency. To couple light from a light-emitting diode (LED) a small gap is left between the waveguide entry surface and the diode, just like coupling light from a laser diode.

The common prism coupler is often used to couple light into a transparent dielectric slab optical waveguide. A prism with refractive index $n_{PRISM} > n_{CLADDING}$ is placed at a distance d_P from an optical waveguide of refractive indices $n_{CORE} > n_{CLADDING}$. An incident wave undergoes total internal reflection at an angle θ_{REFR}. The incident and totally internally reflected waves have a propagation constant $\beta_{PRISM} = n_{PRISM} k_0 \cos(\theta_{PRISM})$. *As the transverse electric field (at the location of the total internal reflection) leaks out of the prism, the waveguide, if sufficiently close (d_P very small), will capture the transverse electric field.* Clearly, the coupling efficiency is correlated to d_P, and thus poor coupling results, if not adjusted carefully. Typically, d_P is the thickness of the waveguide cladding.

If two parallel optical waveguides are very close to each other, then the evanescent transverse distribution of the wave in one waveguide $u_m(y)$ can leak into the nearest neighbor, and over a certain coupling length, the entire wave from the first waveguide leaks into the second and vice versa. This analysis is confined to transparent dielectric slab waveguides. Two parallel waveguides of width d, separation $2a$, and core-cladding refractive indices n_1, n_2 $n_1 > n_2$ are embedded in a material of refractive index slightly lower (weak guidance approximation) than the core and cladding refractive indices. The weak guidance approximation and coupled mode scheme is used to analyze this configuration.

The basic concept of coupled mode theory is that in isolation, the modes in the two waveguides retain their original forms; that is, coupling is a perturbation that alters their amplitudes only, not their transverse spatial distributions. Coupling mode theory is used to determine these modified amplitudes denoted as $a_1(z)$, $a_2(z)$. The coupled first-order differential equations that describe variation of the modified amplitudes along the longitudinal propagation axis z are

$$\frac{d\,a_1(z)}{d\,z} = -jCC_{21}\,e^{j\Delta\beta z}a_2(z) \text{ and } \frac{d\,a_2(z)}{d\,z} = -jCC_{12}\,e^{j\Delta\beta z}a_1(z)$$

where the phase mismatch per unit length is $\Delta\beta = \beta_1 - \beta_2$. The two coupling constants are given by

$$CC_{21} = \frac{(n_2^2 - n^2)k_9^2}{2\beta_1}\int u_1(y)u_2(y)d\,y\ a \leqslant y \leqslant a+d \text{ and } CC_{21}$$

$$= \frac{(n_2^2 - n^2)k_9^2}{2\beta_2}\int u_2(y)u_1(y)d\,y\ -a-d \leqslant y \leqslant -a \qquad (2.49)$$

The constant of proportionality of the variation of $a_1(z)$ is the product of the coupling coefficient and phase mismatch factor $e^{j\Delta\beta z}$.

The differential equations for the amplitudes are solved by invoking the key idea behind coupled modes—a perturbation. So assuming $a_1(z) \neq 0$ $a_2(z) = 0$

$$a_1(z) = a_1(0)e^{-\frac{j\Delta\beta z}{2}}\left(\cos\left(\sqrt{\left(\frac{\Delta\beta}{2}\right)^2 + CC_{12}CC_{21}}\right)z\right.$$

$$\left. -j\left(\frac{\Delta\beta\sin\left(\sqrt{\left(\frac{\Delta\beta}{2}\right)^2 + CC_{12}CC_{21}}\right)z}{2\sqrt{\left(\frac{\Delta\beta}{2}\right)^2 + CC_{12}CC_{21}}}\right)\right)$$

and similarly

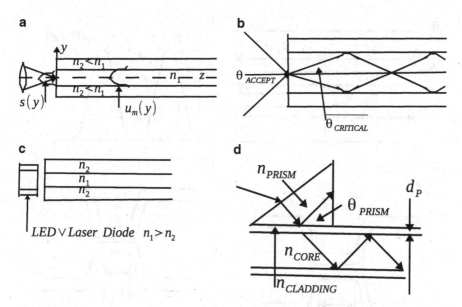

Fig. 2.33 (a) General coupling of light into optical waveguide. (b) Coupling light into multimode optical fiber. (c) Coupling LED/laser diode light into optical fiber. (d) Prism coupling of optical waveguide

$$a_2(z) = -j \frac{a_1(0)CC_{12}}{\sqrt{\left(\left(\frac{\Delta \beta}{2}\right)^2 + CC_{12}CC_2\right)}} \sin\left(\sqrt{\left(\frac{\Delta \beta}{2}\right)^2 + CC_{12}CC_{21}}\, z\right) e^{-j \frac{\Delta \beta}{2} z},$$

$$(2.50)$$

The expressions for signal power in the two waveguides, assuming that the waveguides are made of the same material and have the same refractive index, are $P_1(z) = P_1(0)\left(\cos\left(\sqrt{CC_{12}CC_{21}}\right)z\right)^2$ and $P_1(z) = P_1(0)\left(\sin\left(\sqrt{CC_{12}CC_{21}}\right)z\right)^2$ (Fig. 2.33a, b). The commonly used transparent dielectric slab waveguide directional coupler and its 3 dB variant are shown in Fig. 2.34a, b.

2.15.5   The Directional Coupler

The directional coupler consists of two parallel dielectric optical waveguides, and light is coupled from one waveguide to its neighbor. *The separation between these two parallel waveguides must be very small, so that the exponentially decaying tail of the transverse component of the propagating wave can penetrate the neighboring waveguide.*

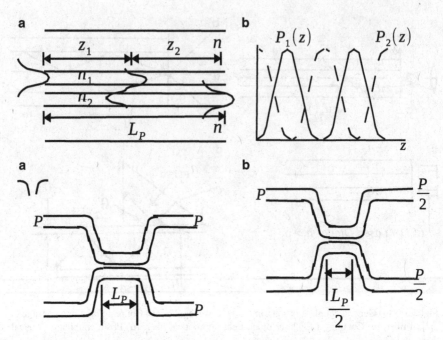

Fig. 2.34 (a) Coupling between transparent dielectric slab waveguides. (b) Periodic exchange of optical signal power between parallel transparent dielectric slab waveguides. (a) Transparent dielectric slab waveguide directional coupler. (b) 3 dB transparent dielectric slab waveguide directional coupler

Consider two dielectric optical waveguides with a distance L over which they are very close to each other, and parallel. The refractive indices of these two waveguides are $n_1^2(x, y) = n_s^2(x, y) + \Delta\, n_1^2(x, y)$ and $n_2^2(x, y) = n_s^2(x, y) + \Delta\, n_2^2(x, y)$. The supported modes are Φ_1 and Φ_2 and the corresponding electric fields are $E_1(x, y, z, t) = A\Phi_1(x, y)e^{-j(\omega t - \beta_1 z)}$ and $E_2(x, y, z, t) = A\Phi_2(x, y)e^{-j(\omega t - \beta_2 z)}$. In the region of very close proximity over a parallel distance L, the modes propagating in the dielectric waveguides obey the following differential equation: $\left[\Delta_T + \frac{\omega^2 n^2(x, y)}{c^2} - \beta^2\right] = 0$ where $n^2(x, y) = n_s^2(x, y) + \Delta\, n_1^2(x, y) + \Delta\, n_2^2(x, y)$ where both $\Delta\, n_1^2(x, y)$ and $\Delta\, n_2^2(x, y)$ are the perturbations to the respective refractive indices of the two waveguides. To solve for the analytical expressions for these coupled modes and the corresponding coupling constants, a few approximations are necessary, to make the solution tractable. As a result of applying those simplifications, the coupling constants, in terms of the modes and refractive indices of the two waveguides, are rewritten as $C_{11} = -\mu\, \omega^2\left[C_{21} A(z)e^{-j\beta_1 z} + C_{22} B(z)e^{-j\beta_2 z}\right]$ where the cross-coupling constant is $C_{21} = \int \Phi_2^{PR}(x, y)\Phi_1(x, y)\Delta\, n_2^2(x, y)dx\, dy$ and $C_{22} = \int \Phi_2^{PR}(x, y)\Phi_2(x, y)\Delta\, n_1^2(x, y)dx\, dy$. The modes in the two waveguides are **not** orthogonal, i.e., $\int \Phi_1^{PR}(x, y)\Phi_2(x, y)dx\, dy \neq 0$. **This allows electromagnetic**

energy to be transferred from one dielectric waveguide to a closely placed adjacent, parallel dielectric waveguide.

The signal power transfer ratio between the two waveguides is given by $T = \frac{\pi^2}{4} sin c^2 \left(\frac{1}{2} \sqrt{1 + \left(\frac{\Delta \beta L_0}{\pi} \right)^2} \right)$. Clearly, the power transfer ratio is dependent on the phase mismatch term $\Delta \beta L_0$, where $L_0 = \frac{\pi}{COUPLING\ CONSTANT}$. T has a maximum value of 1 when $\Delta \beta L_0 = 0$ and has a minimum value of 0 when $\Delta \beta L_0 = \sqrt{3}\ \pi$.

In general, analytical expressions for coupling coefficients are very difficult to derive, and closed-form expressions exist for the lowest order modes, e.g., LP_{01} for which the corresponding coupling coefficient is $C_{01} = \frac{u^2 K_0 \left(\frac{vd}{a} \right)}{k_0 n_1 a^2 V^2 K_1^2(v)}$. Even this is complicated to evaluate numerically, unless the following simplifications are made. $1.5 < V < 2.5$ and $2.0 < \frac{d}{a} < 4.5$ where V, d, and a are, respectively, the optical fiber's V parameter, the separation between the two optical fibers, and the optical fiber radius. Therefore, the coupling coefficient is modified to $C_{01} \simeq \frac{\pi \sqrt{\delta}}{2\ a} e^{- \left(A + \frac{Bd}{a} + \frac{D}{a^2} d^2 \right)}$ where $\delta = \frac{n_1^2 - n_2^2}{n_1^2}$, $A = 5.2789\ V - 3.663\ V + 0.3841\ V^2$, $B = -0.7769 - 1.2252\ V + 0.0152\ V^2$, and $D = -0.0175 - 0.0064 V - 0.009 V^2$. This is the Marcuse approximation.

2.15.5.1   The Practical Optical Waveguide Directional Coupler

The previous analysis of optical fibers overlooks a number of real-world issues related to optical waveguide directional couplers [44]. A working optical fiber directional coupler is made by fusing two parallel fibers while applying a calibrated amount of tension along their lengths. As a result, the diameters of both the optical fibers are reduced in the fused region. Then, the evanescent waves in the fiber cladding are enhanced. As *the core diameters of each of the optical fibers are reduced, larger percentage of the inserted light wave leaks laterally into the corresponding reduced-thickness cladding region. Consequently, their evanescent mode fields start to overlap and interfere, and the optical power can be transferred between the two fibers.* The power transfer coefficient is periodic along the coupling length. The interference between the two evanescent waves is coherent (*in phase*) and wavelength dependent and so the coupling coefficient is wavelength dependent. The degree of wavelength dependency can be controlled and manipulated selecting the coupling length z, so that by modifying the fabrication process, the same fabrication equipment can also be used to make wavelength-selective optical devices, such as wavelength-division multiplexers, demultiplexers, and WDM channel interleavers.

With reference to Fig. 2.35 *power splitting ratio of directional coupler is the ratio of the output of the optical fiber into which the light has been coupled, through the*

Fig. 2.35 Fused fiber
optical directional coupler

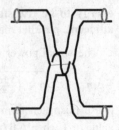

fused region, to the total output power, i.e., *Power Splitting Ratio* $= \alpha =$ $\frac{P_{OUT\ COUPLED}}{P_{OUT\ COUPLED}+P_l}$, and takes into account both absorption and scattering losses. The absorption and scattering losses combine to create a total small loss $\gamma = \frac{P_{OUT\ COUPLED}+P_l}{P_{SOURCE}}$.

The main energy loss mechanism is insertion loss defined as the transmission loss between the input and output. It is dependent on both the directional coupler light energy splitting ratio and excess loss of the fiber, i.e., $LOSS_{INSERTION} = \frac{P_{OUT\ COUPLED}}{P_{SOURCE}} = \alpha\ \gamma$, and the total loss is $LOSS_{TOTAL} = \frac{P_{OUT\ TRANSMITTED}}{P_{SOURCE}} = (1-\alpha)\ \gamma$.

The key coupling ratio is a periodic function of the coupling region and is given as $COUPLING\ RATIO = \left(COUPLING\ RATIO_{MAX}\ \sin\left(\frac{\kappa}{COUPLING\ RATIO_{MAX}}\right)\right)^2 z - z_0$ where $COUPLING\ RATIO_{MAX} \leq 1$ and κ is the periodicity parameter of the coupling region. An empirical expression for κ is $\kappa \approx \frac{21}{a^{\frac{7}{2}}}\lambda^{\frac{5}{2}}$ where a is the radius of the cladding in the fused coupling region, and finally z_0 is the unstretched length of the fused region. The directionality is defined as $D_r = \frac{P_{ret}}{P_{SOURCE}}$, while the reflection is $R_{ref} = \frac{P_{ref}}{P_{SOURCE}}$.

Clearly, there is no well-defined sequence of calculation steps to design and then evaluate the performance characteristics of an optical waveguide directional coupler. The performance characteristics are very sensitive to design parameters in nonlinear ways, so that adjusting some (which are themselves interrelated) does not ensure corresponding change in the performance characteristics. So, like most of the physics, it is back to the experimental table.

2.15.6 Grating Coupler

This type of coupling between propagating [36–47] modes results from periodic perturbation in the structure of the waveguide. There are two types of this coupling—co/contra-directional, i.e., modes propagating in the same or opposite directions. The key propagation equations are

2.15 Coupled Modes Theory and Integrated Optical Devices

$$\frac{d\ A_1(z)}{d\ z} = -j\frac{|\beta_1|}{\beta_1}C_1^m\ _2A_2(z)e^{\ j\Delta\%veta\ z}\ \frac{d\ A_2(z)}{d\ z} = -j\frac{|\beta_2|}{\beta_2}C_2^m\ _1A_1(z)e^{-j\ \Delta\ \%veta\ z}$$

where $\Delta\beta = \beta_1 - \beta_2 + \frac{2\%oi\ n}{\Gamma}$ and $C_k^m\ _l = \frac{\omega}{4}\int E_l^{PR}|\epsilon_m\ E_m|dx\ dy$.

In the codirectional propagation case, $\frac{|\beta_1|}{\beta_1} = \frac{|\beta_2|}{\beta_2}$ and with the help of two temporary variables $\kappa = -jC_{12}^m$ and $K = |\kappa^2| + \Delta\ \beta^2$ the solutions for the amplitude become $A_1(z) = e^{\ j\Delta\beta z}\left(A_1(0)\left(\cos(K\ z) + j\frac{\Delta\beta}{K}\sin(K\ z)\right) - A_2(0)\frac{\kappa}{K^{PR}}\sin(K\ z)\right)$ and $A_2(z) = e^{-j\Delta\beta z}\left(A_2(0)\left(\cos(K\ z) + j\frac{\Delta\beta}{K}\sin(K\ z)\right) - A_1(0)\frac{\kappa}{K^{PR}}\sin(K\ z)\right)$. Now for resonant coupling $\Delta\ \beta = 0$ and so the solutions simplify to $A_1(z) = A_1(0)$ cos $(K\ z)$ and $A_2(z) = A_1(0)\frac{\kappa^{PR}}{|\kappa|}\sin(|\kappa|)z$. The interacting modes exchange energy in the interaction length $\frac{\pi}{K}$, satisfying $\frac{d}{dz}\left(|A_1{}^2| + |A_2{}^2|\right) = 0$.

In contra-directional coupling, two modes travel in opposite directions in a single-mode optical waveguide. This coupling depends on the refractive index of the core of the optical waveguide being modulated, i.e., Bragg grating: $n(z) = n_{CORE} + \Delta n$ sin $\left(\frac{2\pi}{\Gamma}\right)z$. In this case, $\frac{|\beta_1|}{\beta_1} = 1$, $\frac{|\beta_2|}{\beta_2} = -1$ and $\Delta\ \beta = |\beta_1| + |\beta_1| - \frac{2\pi}{\Gamma} = 2k_0\ n_{eff} - \frac{2\pi}{\Gamma}$. In these expressions, Γ is the grating period. The expressions for the amplitudes are $A_1(z) = \frac{A\cosh(\kappa(z-L))}{\cosh(z\ L)}$ and $A_2(z) = \frac{A\sinh(\kappa(z-L))}{\cosh(z\ L)}$ where $\kappa = \frac{\pi\Delta n}{\lambda_0}$. In effect, the periodic structure acts as a mirror with a reflection coefficient $R = (\tanh(\kappa\ L))^\square$.

In summary, this chapter examines in minute detail how light propagates through optical waveguides (fiber, slab). The analysis draws upon geometric optics concepts (reflection, refraction, critical angle, and total internal reflection), and uses a rigorous mathematical approach based on Maxwell's equations of electromagnetism applied to dielectric media. Specifically mode propagation through optical fibers and dielectric slab waveguides is examined in detail (allowed modes, their propagation constants and group velocities, etc.). The issues of signal energy loss due to attenuation, absorption, scattering, and bending and the methods to counter these problems (fiber optical laser) are presented and its working from a quantum mechanical point of view is examined in detail.

Coupled mode theory that explains how propagating modes in the same/neighboring dielectric optical waveguides (fiber, slab) interact is then examined in detail and then two real-world optical signal processing devices, the directional and grating couplers, are examined.

The next chapter examines in detail how a set of supplied C computer language executables (for both the popular Linux and Windows operating systems) are used to design and analyze the properties of dielectric optical waveguides (optical fiber, slab) and estimate the parameters of directional and grating couplers.

References

1. http://www-eng.lbl.gov/~shuman/NEXT/CURRENT_DESIGN/TP/FO/Lect4-Optical%20 waveguides.pdf
2. https://courses.cit.cornell.edu/ece533/Lectures/handout8.pdf
3. https://www.fiberoptics4sale.com/blogs/wave-optics/symmetric-slab-waveguides
4. http://web.mit.edu/6.013_book/www/chapter13/13.5.html
5. http://www.phy.cuhk.edu.hk/~jfwang/PDF/Chapter%203%20waveguide.pdf
6. http://users.ntua.gr/eglytsis/IO/Slab_Waveguides_Chapter.pdf
7. Saleh Bahaa EA, Teich MC (1991) Fundamentals of photonics. John Wiley & Sons, Inc. ISBNs: 0-471-83965-5 (Hardback); 0-471-2-1374-8 (Electronic)
8. http://courses.washington.edu/me557/sensors/waveguide.pdf
9. http://pongsak.ee.engr.tu.ac.th/le426/doc/slab_waveguide.pdf
10. https://www.osapublishing.org/aop/fulltext.cfm?uri=aop-1-1-58&id=176222
11. https://faculty.kfupm.edu.sa/ee/ajmal/files/Ajmal_Thesis_Chap2.pdf
12. http://www.faculty.jacobs-university.de/dknipp/c320352/lectures/3%20waveguides.pdf
13. https://link.springer.com/chapter/10.1007%2F978-1-4615-3106-7_2
14. https://link.springer.com/chapter/10.1007%2F978-0-387-73068-4_2
15. https://www.researchgate.net/publication/234932545_Analysis_of_TE_Transverse_Electric_ modes_of_symmetric_slab_waveguide
16. http://www.m-hikari.com/astp/astp2012/astp25-28-2012/ramzaASTP25-28-2012.pdf
17. http://opticaapplicata.pwr.edu.pl/files/pdf/2015/no3/optappl_4503p405.pdf
18. https://arxiv.org/pdf/1511.03479.pdf
19. https://www.osapublishing.org/josa/abstract.cfm?uri=josa-51-5-491
20. http://users.ntua.gr/eglytsis/IO/Cylindrical_Waveguides_p.pdf
21. Kapoor A, Singh GS (2000) Mode classification in cylindrical dielectric waveguides. J Lightwave Technol 18(6)
22. https://www.sciencedirect.com/topics/physics-and-astronomy/dielectric-waveguides
23. https://link.springer.com/chapter/10.1007/978-0-387-49799-0_5
24. https://www.researchgate.net/publication/232986168_Single-mode_Anisotropic_Cylindrical_ Dielectric_Waveguides
25. Yu E, Smol'kin, Valovik DV (2015) Guided electromagnetic waves propagating in a two-layer cylindrical dielectric waveguide with inhomogeneous nonlinear permittivity. Adv Math Phys:614976. https://doi.org/10.1155/2015/614976
26. Novotny L, Hafner C (1994) Light propagation in a cylindrical waveguide with complex, metallic, dielectric function. Phys Rev E50:4094
27. Bruno WM, Bridfes WB (1988) Flexible dielectric waveguides with powder cores. IEEE Trans Microwave Theory Tech 36(5)
28. Taylor HF, Yariv A (1974) Guided wave optics. Proc IEEE 62(8)
29. https://www.ad-net.com.tw/optical-power-loss-attenuation-in-fiber-access-types-values-and-sources/
30. https://www.ppc-online.com/blog/understanding-optical-loss-in-fiber-networks-and-how-to-tackle-it
31. https://www.juniper.net/documentation/en_US/release-independent/junos/topics/concept/fiber-optic-cable-signal-loss-attenuation-dispersion-understanding.html
32. https://www.fiberoptics4sale.com/blogs/archive-posts/95048006-optical-fiber-loss-and-attenuation
33. https://www.sciencedirect.com/topics/engineering/fiber-loss
34. https://community.fs.com/blog/understanding-loss-in-fiber-optic.html
35. https://www.elprocus.com/attenuation-in-optical-fibre-types-and-its-causes/
36. https://www.cablesys.com/updates/fiber-loss-in-fiber-optic-performance/
37. https://www.thefoa.org/tech/ref/testing/test/dB.html
38. http://opti500.cian-erc.org/opti500/pdf/sm/Module3%20Optical%20Attenuation.pdf

39. https://wiki.metropolia.fi/display/Physics/Attenuation+in+Fiber+Optics
40. http://www.dsod.p.lodz.pl/materials/OPT04_IFE.pdf
41. http://electron9.phys.utk.edu/optics421/modules/m8/optical_fiber_measurements.htm
42. https://rmd.ac.in/dept/ece/Supporting_Online_%20Materials/7/OCN/unit2.pdf
43. https://www.rp-photonics.com/passive_fiber_optics7.html
44. https://www.sciencedirect.com/topics/engineering/fiber-directional-coupler
45. https://link.springer.com/chapter/10.1007%2F978-1-4757-1177-6_19
46. https://www.wolfram.com/mathematica/
47. https://www.mathworks.com/products/matlab.html

Chapter 3
Analysis and Design Examples of Optical Fibers, Optical Slab Waveguides

In the following sections, all computer-generated text is in bold font and the ready-to-use C computer language executables that perform the analysis and design calculations use the universally accepted MKS system of units. Each of these ready-to-use executables preferably use the command line input, as this guarantees that the input parameters are entered in the correct order; for example if the values of the core and cladding refractive indices get interchanged by mistake, the computed results will be totally incorrect.

3.1 Simple-Mode Analysis of Step Index Multimode Optical Fiber

The ready-to-use C computer language [1] executable **wvgdcsi** (versions available for both the popular Linux and Windows operating systems) exploits the normalized longitudinal wave number concept to compute the possible longitudinal wavenumbers, core, and cladding parameters of a step index multimode optical fiber, under the weak guidance approximation. Unfortunately, the corresponding mode numbers(l, m) cannot be determined, using this method. Typing `./wvgdcsi` at the Linux shell command prompt generates the help information:

```
$ ./wvgdcsi
incorrect|insufficient parameters
```

The original version of this chapter was revised. The correction to this chapter is available at https://doi.org/10.1007/978-3-030-93631-0_4

Supplementary Information The online version contains supplementary material available at (https://doi.org/10.1007/978-3-030-93631-0_3).

A. Banerjee, *Optical Waveguides Analysis and Design*, https://doi.org/10.1007/978-3-030-93631-0_3

```
interactive
./wvgdcsi i|I
batch|command line argument
./wvgdcsi b|B|c|C
<frequency THz>
<refractive index core>
<refractive index cladding>
<maximum radius micron>
<increment value for propagation constant estimation>
sample batch|command line argument input
./wvgdcsi b|B|c|C 500 1.5 1.475 35 0.05
```

Using the sample command line argument input, the generated results are:

```
$ ./wvgdcsi b 500 1.5 1.475 35 0.05
estimated possible beta core cladding parameters
6.285365e+00   97.376   22.340
6.290725e+00   94.779   31.593
6.296081e+00   92.108   38.693
6.301432e+00   89.358   44.679
6.306779e+00   86.521   49.953
6.312121e+00   83.587   54.721
6.317458e+00   80.546   59.105
6.322791e+00   77.387   63.186
6.328120e+00   74.092   67.019
6.333444e+00   70.644   70.644
6.338764e+00   67.019   74.092
6.344079e+00   63.186   77.387
6.349389e+00   59.105   80.546
6.354696e+00   54.721   83.587
6.359998e+00   49.953   86.521
6.365295e+00   44.679   89.358
6.370588e+00   38.693   92.108
6.375877e+00   31.593   94.779
6.381161e+00   22.340   97.376
step index fiber weak guiding
free space wavelength 6.000000e-07 m
numerical aperture 0.27272
V parameter 99.906
acceptance angle 15.8341 degree
critical angle 79.5650 degree
estimated max. modes 4049.306
approximate cut-off frequency 3.282530e+12 Hz
```

As mentioned earlier, although this simple scheme allows estimation of the longitudinal propagation constants (beta) of possible propagating modes, the corresponding mode numbers cannot be estimated. This drawback is addressed using the advanced/improved scheme examined next, applicable to both step and graded index fibers.

3.2   Analysis and Design of Step and Graded Index Optical Fiber at 1500 nm and 1300 nm

The supplied ready-to-use C computer language [1] executable **wvgdesof** accepts key optical fiber input parameters as:

- Core, cladding refractive index
- Wavelength of the input light
- Core radius whether the fiber is step or graded index

It computes essential optical fiber parameters as:

- Acceptance angle
- Numerical aperture
- V parameter
- Maximum possible number of allowed modes
- Cutoff frequency and wavelength
- Required radius of the single-mode fiber
- The longitudinal propagation constants of the first few important linearized modes and their normalized propagation constants, core parameters, and group velocities

Help information is readily available by typing **./wvgdesof** at the Linux shell command prompt, which generates:

```
$ ./wvgdesof
incorrect|insufficient arguments
interactive
./wvgdesof i|I
batch|command line argument
./wvgdesof b|B|c|C
<core radius micron>
<refractive index core center>
<refractive index cladding>
<wavelength nanometer>
<graded index profile parameter - NONZERO FOR GRADED INDEX FIBER ONLY
<decrement value for single mode fiber radius estimation micron>
sample command line input step index
./wvgdesof b 10 1.5 1.48 1500 0 0.2
sample command line input graded index
./wvgdesof b 15 1.5 1.48 1300 2 0.3
```

Using the supplied command line argument list for the step index fiber

```
./wvgdesof b 10 1.5 1.48 1500 0 0.2
```

generates:

```
step index modes (1 m) prop. constants norm. prop. constants core param
group velocity (m/s)
01   9.408225e+06  0.906    3.139    2.996250e+08
```

```
02   9.372900e+06   0.623    6.272    2.985000e+08
03   9.314025e+06   0.155    9.393    2.966250e+08
11   9.393506e+06   0.788    4.707    2.991562e+08
12   9.346406e+06   0.412    7.835    2.976562e+08
13   9.275756e+06  -0.147   10.948    2.954062e+08
21   9.372900e+06   0.623    6.272    2.985000e+08
22   9.314025e+06   0.155    9.393    2.966250e+08
23   9.231600e+06  -0.495   12.497    2.940000e+08
31   9.346406e+06   0.412    7.835    2.976562e+08
32   9.275756e+06  -0.147   10.948    2.954062e+08
33   9.181556e+06  -0.887   14.040    2.924062e+08
numerical aperture 0.2441
critical angle 80.674 degree
acceptance angle 14.138 degree
V parameter 10.2210
step index max. possible number of modes 42.38
cut off wavelength 6.374817e-06 m
cut off frequency 4.706018e+13 Hz
estimated single mode radius 2.0000 micron
```

For the generated data shown in tabular format above, the first column represents the mode numbers (l, m) of the linearly polarized modes, the second column shows the longitudinal propagation constant for each (l, m) mode, the third column represents the normalized propagation constant, the fourth column indicates the core parameter, and finally the last column represents the group velocity (m/s) for each mode.

The numbers in the first column of the output are the mode numbers (l, m) of the first few allowed linearized modes through the step index fiber. As the V parameter is much greater than 2.405, it is a multimode fiber and the maximum possible number of modes is 42. Consequently, to convert the fiber to a single-mode optical fiber, its radius must be reduced to 2.0 micron. In addition, the lowest possible frequency that can propagate through the fiber is 37 THz. *Therefore, by appropriately adjusting the input parameters as core, cladding refractive indices, and core radius, it is possible to design an optical fiber that can have specific values for numerical aperture, acceptance angle, critical angle, and V parameter, for a given input light wavelength.* For the step index fiber, the core index profile parameter is ∞ for computer program, with the input as zero.

The sample command line argument list for the graded index fiber with a common index profile parameter of 2 is used to analyze and design a graded index fiber.

```
./wvgdesof C 15 1.5 1.48 1300 2 0.3

graded index possible modes propagation constants group velocities (m/
s)
0   7.246154e+06   3.000000e+08
1   7.235304e+06   3.000000e+08
2   7.230809e+06   3.000000e+08
3   7.227361e+06   3.000000e+08
4   7.224453e+06   3.000000e+08
numerical aperture 0.2441
critical angle 80.674 degree
acceptance angle 14.138 degree
V parameter 17.6901
```

```
graded index max. possible number of modes 78.24
cut off wavelength 9.562225e-06 m
cut off frequency 3.137345e+13 Hz
estimated single mode radius 1.5000 micron
```

It is left to the reader to analyze the above results generated by **wvgdesof** for the case of the graded index fiber, and extract meaningful insights into the behavior and performance characteristics of this fiber.

3.3   Allowed Propagating Mode Estimation in Slab Dielectric Waveguide Using Simple Method A (Sect. 2.9)

The simple scheme to calculate the transverse propagation constants in a slab dielectric waveguide has been implemented in a C computer language [1] executable **wvgdseigeneng.** *Although this method suffers from the drawback that mode numbers cannot be computed, it clearly demonstrates how the transcendental eigenvalue equation for both the even and odd modes for both TE and TM modes may be solved numerically in a straightforward manner.* Slab dielectric waveguides do not support linearized or quasi-TEM modes, only TE or TM. Also, because of their short lengths (unlike optical fibers, slab dielectric waveguides are designed to process optical signals—do not carry signals over large distances), signal attenuation, signal power absorption, and dispersion are negligible and can be safely ignored. Moreover, instead of the core and cladding refractive indices as input, dielectric constants are used in the transcendental eigenvalue equations. **wvgdseigeneng**, available for both the Linux and Windows operating systems, is typically invoked using command line arguments—help information is obtained simply by typing at the command prompt:

```
$ ./wvgdseigeneng
incorrect|insufficient arguments
interactive
./wvgdseigeneng i|I
batch|command line argument
./wvgdseigeneng b|B|c|C
TE|TM switch 1|2
<even|odd switch 1|2>
<dielectric constant slab>
<dielectric constant slab substrate-top>
<frequency THz>
<slab thickness micron>
<tolerance on comparison>
<increment for iteration>
sample command line input TE even
./wvgdseigeneng b 1 1 2.25 2.1025 560 2 0.96 1.0
sample command line input TE odd
./wvgdseigeneng B 1 2 2.25 2.1025 700 2 0.96 1.0
sample command line input TM even
```

```
./wvgdseigeneng c 2 1 2.25 2.1025 650 2 0.96 1.0
sample command line input TM odd
./wvgdseigeneng B 2 2 2.25 2.1025 750 2 0.96 1.0
```

So, using the sample command line argument for input for even TE mode generates:

```
$ ./wvgdseigeneng b 1 1 2.25 2.1025 560 2 0.96 1.0
TE even transverse wave number estimated value 2759494.381
```

Similarly, for the TE odd mode, the results obtained by using the sample command line argument input are:

```
./wvgdseigeneng B 1 2 2.25 2.1025 700 2 0.96 1.0
TE odd transverse wave number estimated value 3726340.966
```

And the sample command line argument input for the TM even mode gives:

```
$ ./wvgdseigeneng c 2 1 2.25 2.1025 650 2 0.96 1.0
TM even transverse wave number estimated value 3047530.463
```

The TM odd mode is left as an exercise for the reader.

3.4   Allowed Propagating Mode Estimation in Slab Dielectric Waveguide Using Simple Method B (Sect. 2.10)

The slab dielectric waveguide design example (in Sect. 3.2) accurately calculates the transverse (perpendicular to the wave propagation axis) propagation constant for each of the even/odd TE/TM modes, *but cannot estimate the corresponding mode numbers*. **The slab dielectric waveguide propagating mode eigenvalue determination scheme examined in Sect. 2.10 can estimate the even/odd TE/TM mode number and corresponding eigenvalue.**

The C computer language [1] executable **wcgdseigenA** available for both the Linux and Windows operating systems implements the scheme examined in detail in Sect. 2.10. Typing ./wvgdseigenB at the Linux shell command prompt generates the help information:

```
$ ./wvgdseigenA
incorrect|insufficient arguments
interactive
./wvgdseigenA i|I
batch|command line argument
./wvgdseigenA b|B|c|C
```

```
<even|odd 1|2>
<TE|TM 1|2>
<dielectric constant slab>
<dielectric constant slab substrate top>
<frequency THz>
<slab thickness micron>
<tolerance on comparison>
<increment for iteration>
sample command line input TE even
./wvgdseigenA b 1 1 2.25 2.1025 500 10 0.96 1.0
sample command line input TE odd
./wvgdseigenA B 2 1 2.25 2.1025 375 15 0.96 1.0
sample command line input TM even
./wvgdseigenA c 1 2 2.25 2.1025 750 25 0.96 1.0
sample command line input TM odd
./wvgdseigenA C 2 2 2.25 2.1025 1000 30 0.96 1.0
```

Although it can be executed in both the interactive and batch/command line argument modes, the batch/command line argument mode is preferable. Using the supplied sample command line argument input for even TE mode, the following results are generated:

```
./wvgdseigenA b 1 1 2.25 2.1025 500 10 0.96 1.0
even TE eigenvalue
m 2   kxd 4.490   LHS 4.365230e+00   RHS 4.422250e+00
even TE eigenvalue
m 4   kxd 7.480   LHS 2.495260e+00   RHS 2.548089e+00
even TE eigenvalue
m 8   kxd 13.410  LHS 1.117294e+00   RHS 1.123812e+00
```

In the above-generated output, consider:

```
m 2   kxd 4.490   LHS 4.365230e+00   RHS 4.422250e+00
```

- "m 2" indicates the mode number.
- "kxd 4.490" indicates the eigenvalue, where "kxd" is the product of the transverse wave number and the waveguide thickness. The transverse wave number is calculated by dividing this product with the known value of the thickness.

The numbers labeled as LHS and RHS are the numerical values for the left- and right-hand sides, respectively, for the transcendental eigenvalue equation being solved, for the generated eigenvalue.

The TE odd mode eigenvalues are generated by using the appropriate supplied sample command line arguments:

```
$ ./wvgdseigenA B 2 1 2.25 2.1025 375 15 0.96 1.0
odd TE eigenvalue
m 3    kxd 6.010   LHS 3.628630e+00   RHS 3.569001e+00
odd TE eigenvalue
m 5    kxd 9.025   LHS 2.298365e+00   RHS 2.366687e+00
odd TE eigenvalue
m 7    kxd 12.015  LHS 1.595208e+00   RHS 1.626036e+00
odd TE eigenvalue
m 9    kxd 14.980  LHS 1.131526e+00   RHS 1.122012e+00
```

In a similar fashion, the TM even eigenvalues are generated using the appropriate supplied sample command line arguments:

```
./wvgdseigenA c 1 2 2.25 2.1025 750 25 0.96 1.0
even TM eigenvalue
m 8    kxd 13.960  LHS 5.680403e+00   RHS 5.585214e+00
```

Finally, the TM odd eigenvalue(s) can be generated using the appropriate supplied sample command line arguments:

```
$ ./wvgdseigenA C 2 2 2.25 2.1025 1000 30 0.96 1.0
odd TM eigenvalue
m 3    kxd 6.235   LHS 2.067954e+01   RHS 2.073715e+01
```

3.5   Designing Optical Slab Waveguides and Generating Its Mode Charts for TE Modes

The ready-to-use C computer language [1] executable **wvgdes** implements the simple dielectric slab waveguide technique examined in Sect. 2.11. Typing ./wvgdes at the Linux shell command prompt generates the following help information:

```
$ ./wvgdes
incorrect | insufficient arguments
interactive
./wvgdes i | I
batch | command line argument
./wvgdes b | B | c | C
<frequency THz>
<optical fiber | slab waveguide diameter | height micron>
<refractive index core>
<refractive index cladding>
sample batch | command line argument input
./wvgdes b | B | c | C 350 35 3.6 3.55
```

Using the supplied sample batch/command line argument input generates the following output:

```
mode 0.000
80.480   147.000
81.480   147.000
82.480   147.000
83.480   147.000
84.480   147.000
85.480   147.000
86.480   147.000
87.480   147.000
88.480   147.000
89.480   147.000
mode 1.000
80.480   147.836
81.480   147.933
82.480   148.055
83.480   148.215
84.480   148.433
85.480   148.746
86.480   149.235
87.480   150.105
88.480   152.087
89.480   161.086
mode 2.000
80.480   148.672
81.480   148.866
82.480   149.111
83.480   149.431
84.480   149.866
85.480   150.492
86.480   151.469
87.480   153.210
88.480   157.175
89.480   175.172
mode 3.000
80.480   149.509
81.480   149.799
82.480   150.166
83.480   150.646
84.480   151.299
85.480   152.238
86.480   153.704
87.480   156.314
88.480   162.262
89.480   189.258
mode 4.000
80.480   150.345
81.480   150.732
82.480   151.222
83.480   151.861
84.480   152.732
85.480   153.983
86.480   155.938
87.480   159.419
```

```
88.480   167.349
89.480   203.344
mode 5.000
80.480   151.181
81.480   151.665
82.480   152.277
83.480   153.077
84.480   154.164
85.480   155.729
86.480   158.173
87.480   162.524
88.480   172.436
89.480   217.430
mode 6.000
80.480   152.017
81.480   152.598
82.480   153.333
83.480   154.292
84.480   155.597
85.480   157.475
86.480   160.408
87.480   165.629
88.480   177.524
89.480   231.516
mode 7.000
80.480   152.854
81.480   153.531
82.480   154.388
83.480   155.508
84.480   157.030
85.480   159.221
86.480   162.642
87.480   168.733
88.480   182.611
89.480   245.603
mode 8.000
80.480   153.690
81.480   154.464
82.480   155.444
83.480   156.723
84.480   158.463
85.480   160.967
86.480   164.877
87.480   171.838
88.480   187.698
89.480   259.689
mode 9.000
80.480   154.526
81.480   155.397
82.480   156.499
83.480   157.938
84.480   159.896
85.480   162.713
```

```
86.480   167.112
87.480   174.943
88.480   192.785
89.480   273.775
mode 10.000
80.480   155.362
81.480   156.330
82.480   157.555
83.480   159.154
84.480   161.329
85.480   164.458
86.480   169.346
87.480   178.048
88.480   197.873
89.480   287.861
mode 11.000
80.480   156.199
81.480   157.263
82.480   158.610
83.480   160.369
84.480   162.762
85.480   166.204
86.480   171.581
87.480   181.152
88.480   202.960
89.480   301.947
mode 12.000
80.480   157.035
81.480   158.196
82.480   159.666
83.480   161.584
84.480   164.195
85.480   167.950
86.480   173.815
87.480   184.257
88.480   208.047
89.480   316.033
mode 13.000
80.480   157.871
81.480   159.129
82.480   160.721
83.480   162.800
84.480   165.627
85.480   169.696
86.480   176.050
87.480   187.362
88.480   213.135
89.480   330.119
mode 14.000
80.480   158.707
81.480   160.062
82.480   161.777
83.480   164.015
```

```
84.480    167.060
85.480    171.442
86.480    178.285
87.480    190.467
88.480    218.222
89.480    344.205
mode 15.000
80.480    159.544
81.480    160.995
82.480    162.832
83.480    165.231
84.480    168.493
85.480    173.188
86.480    180.519
87.480    193.571
88.480    223.309
89.480    358.291
mode 16.000
80.480    160.380
81.480    161.928
82.480    163.888
83.480    166.446
84.480    169.926
85.480    174.934
86.480    182.754
87.480    196.676
88.480    228.396
89.480    372.377
mode 17.000
80.480    161.216
81.480    162.861
82.480    164.943
83.480    167.661
84.480    171.359
85.480    176.679
86.480    184.988
87.480    199.781
88.480    233.484
89.480    386.463
mode 18.000
80.480    162.052
81.480    163.794
82.480    165.999
83.480    168.877
84.480    172.792
85.480    178.425
86.480    187.223
87.480    202.886
88.480    238.571
89.480    400.549
mode 19.000
80.480    162.889
81.480    164.727
```

```
82.480   167.054
83.480   170.092
84.480   174.225
85.480   180.171
86.480   189.458
87.480   205.991
88.480   243.658
89.480   414.635
mode 20.000
80.480   163.725
81.480   165.660
82.480   168.110
83.480   171.307
84.480   175.658
85.480   181.917
86.480   191.692
87.480   209.095
88.480   248.746
89.480   428.721
mode 21.000
80.480   164.561
81.480   166.593
82.480   169.165
83.480   172.523
84.480   177.090
85.480   183.663
86.480   193.927
87.480   212.200
88.480   253.833
89.480   442.808
mode 22.000
80.480   165.397
81.480   167.526
82.480   170.221
83.480   173.738
84.480   178.523
85.480   185.409
86.480   196.162
87.480   215.305
88.480   258.920
89.480   456.894
mode 23.000
80.480   166.234
81.480   168.459
82.480   171.276
83.480   174.954
84.480   179.956
85.480   187.155
86.480   198.396
87.480   218.410
88.480   264.007
89.480   470.980
```

mode 24.000
80.480   167.070
81.480   169.392
82.480   172.331
83.480   176.169
84.480   181.389
85.480   188.900
86.480   200.631
87.480   221.514
88.480   269.095
89.480   485.066
mode 25.000
80.480   167.906
81.480   170.325
82.480   173.387
83.480   177.384
84.480   182.822
85.480   190.646
86.480   202.865
87.480   224.619
88.480   274.182
89.480   499.152
mode 26.000
80.480   168.742
81.480   171.258
82.480   174.442
83.480   178.600
84.480   184.255
85.480   192.392
86.480   205.100
87.480   227.724
88.480   279.269
89.480   513.238
mode 27.000
80.480   169.579
81.480   172.191
82.480   175.498
83.480   179.815
84.480   185.688
85.480   194.138
86.480   207.335
87.480   230.829
88.480   284.356
89.480   527.324
mode 28.000
80.480   170.415
81.480   173.124
82.480   176.553
83.480   181.030
84.480   187.121
85.480   195.884
86.480   209.569
87.480   233.933

```
88.480    289.444
89.480    541.410
mode 29.000
80.480    171.251
81.480    174.057
82.480    177.609
83.480    182.246
84.480    188.553
85.480    197.630
86.480    211.804
87.480    237.038
88.480    294.531
89.480    555.496
mode 30.000
80.480    172.087
81.480    174.990
82.480    178.664
83.480    183.461
84.480    189.986
85.480    199.375
86.480    214.038
87.480    240.143
88.480    299.618
89.480    569.582
mode 31.000
80.480    172.924
81.480    175.923
82.480    179.720
83.480    184.677
84.480    191.419
85.480    201.121
86.480    216.273
87.480    243.248
88.480    304.706
89.480    583.668
mode 32.000
80.480    173.760
81.480    176.856
82.480    180.775
83.480    185.892
84.480    192.852
85.480    202.867
86.480    218.508
87.480    246.352
88.480    309.793
89.480    597.754
mode 33.000
80.480    174.596
81.480    177.789
82.480    181.831
83.480    187.107
84.480    194.285
85.480    204.613
```

```
86.480    220.742
87.480    249.457
88.480    314.880
89.480    611.840
mode 34.000
80.480    175.432
81.480    178.722
82.480    182.886
83.480    188.323
84.480    195.718
85.480    206.359
86.480    222.977
87.480    252.562
88.480    319.967
89.480    625.927
mode 35.000
80.480    176.268
81.480    179.656
82.480    183.942
83.480    189.538
84.480    197.151
85.480    208.105
86.480    225.211
87.480    255.667
88.480    325.055
89.480    640.013
mode 36.000
80.480    177.105
81.480    180.589
82.480    184.997
83.480    190.753
84.480    198.584
85.480    209.851
86.480    227.446
87.480    258.771
88.480    330.142
89.480    654.099
mode 37.000
80.480    177.941
81.480    181.522
82.480    186.053
83.480    191.969
84.480    200.017
85.480    211.596
86.480    229.681
87.480    261.876
88.480    335.229
89.480    668.185
mode 38.000
80.480    178.777
81.480    182.455
82.480    187.108
83.480    193.184
```

```
84.480   201.449
85.480   213.342
86.480   231.915
87.480   264.981
88.480   340.317
89.480   682.271
mode 39.000
80.480   179.613
81.480   183.388
82.480   188.164
83.480   194.400
84.480   202.882
85.480   215.088
86.480   234.150
87.480   268.086
88.480   345.404
89.480   696.357
mode 40.000
80.480   180.450
81.480   184.321
82.480   189.219
83.480   195.615
84.480   204.315
85.480   216.834
86.480   236.385
87.480   271.191
88.480   350.491
89.480   710.443
mode 41.000
80.480   181.286
81.480   185.254
82.480   190.275
83.480   196.830
84.480   205.748
85.480   218.580
86.480   238.619
87.480   274.295
88.480   355.578
89.480   724.529
mode 42.000
80.480   182.122
81.480   186.187
82.480   191.330
83.480   198.046
84.480   207.181
85.480   220.326
86.480   240.854
87.480   277.400
88.480   360.666
89.480   738.615
mode 43.000
80.480   182.958
81.480   187.120
```

```
82.480    192.386
83.480    199.261
84.480    208.614
85.480    222.071
86.480    243.088
87.480    280.505
88.480    365.753
89.480    752.701
mode 44.000
80.480    183.795
81.480    188.053
82.480    193.441
83.480    200.476
84.480    210.047
85.480    223.817
86.480    245.323
87.480    283.610
88.480    370.840
89.480    766.787
mode 45.000
80.480    184.631
81.480    188.986
82.480    194.497
83.480    201.692
84.480    211.480
85.480    225.563
86.480    247.558
87.480    286.714
88.480    375.927
89.480    780.873
mode 46.000
80.480    185.467
81.480    189.919
82.480    195.552
83.480    202.907
84.480    212.912
85.480    227.309
86.480    249.792
87.480    289.819
88.480    381.015
89.480    794.959
mode 47.000
80.480    186.303
81.480    190.852
82.480    196.607
83.480    204.123
84.480    214.345
85.480    229.055
86.480    252.027
87.480    292.924
88.480    386.102
89.480    809.046
mode 48.000
```

```
80.480    187.140
81.480    191.785
82.480    197.663
83.480    205.338
84.480    215.778
85.480    230.801
86.480    254.261
87.480    296.029
88.480    391.189
89.480    823.132
mode 49.000
80.480    187.976
81.480    192.718
82.480    198.718
83.480    206.553
84.480    217.211
85.480    232.547
86.480    256.496
87.480    299.133
88.480    396.277
89.480    837.218
mode 50.000
80.480    188.812
81.480    193.651
82.480    199.774
83.480    207.769
84.480    218.644
85.480    234.292
86.480    258.731
87.480    302.238
88.480    401.364
89.480    851.304
mode 51.000
80.480    189.648
81.480    194.584
82.480    200.829
83.480    208.984
84.480    220.077
85.480    236.038
86.480    260.965
87.480    305.343
88.480    406.451
89.480    865.390
mode 52.000
80.480    190.485
81.480    195.517
82.480    201.885
83.480    210.199
84.480    221.510
85.480    237.784
86.480    263.200
87.480    308.448
88.480    411.538
```

```
89.480   879.476
mode 53.000
80.480   191.321
81.480   196.450
82.480   202.940
83.480   211.415
84.480   222.943
85.480   239.530
86.480   265.435
87.480   311.552
88.480   416.626
89.480   893.562
mode 54.000
80.480   192.157
81.480   197.383
82.480   203.996
83.480   212.630
84.480   224.375
85.480   241.276
86.480   267.669
87.480   314.657
88.480   421.713
89.480   907.648
mode 55.000
80.480   192.993
81.480   198.316
82.480   205.051
83.480   213.846
84.480   225.808
85.480   243.022
86.480   269.904
87.480   317.762
88.480   426.800
89.480   921.734
mode 56.000
80.480   193.830
81.480   199.249
82.480   206.107
83.480   215.061
84.480   227.241
85.480   244.768
86.480   272.138
87.480   320.867
88.480   431.888
89.480   935.820
mode 57.000
80.480   194.666
81.480   200.182
82.480   207.162
83.480   216.276
84.480   228.674
85.480   246.513
86.480   274.373
```

```
87.480    323.972
88.480    436.975
89.480    949.906
mode 58.000
80.480    195.502
81.480    201.115
82.480    208.218
83.480    217.492
84.480    230.107
85.480    248.259
86.480    276.608
87.480    327.076
88.480    442.062
89.480    963.992
mode 59.000
80.480    196.338
81.480    202.048
82.480    209.273
83.480    218.707
84.480    231.540
85.480    250.005
86.480    278.842
87.480    330.181
88.480    447.149
89.480    978.078
mode 60.000
80.480    197.175
81.480    202.981
82.480    210.329
83.480    219.922
84.480    232.973
85.480    251.751
86.480    281.077
87.480    333.286
88.480    452.237
89.480    992.164
mode 61.000
80.480    198.011
81.480    203.914
82.480    211.384
83.480    221.138
84.480    234.406
85.480    253.497
86.480    283.311
87.480    336.391
88.480    457.324
89.480   1006.251
mode 62.000
80.480    198.847
81.480    204.847
82.480    212.440
83.480    222.353
84.480    235.838
```

```
85.480   255.243
86.480   285.546
87.480   339.495
88.480   462.411
89.480   1020.337
mode 63.000
80.480   199.683
81.480   205.780
82.480   213.495
83.480   223.569
84.480   237.271
85.480   256.988
86.480   287.781
87.480   342.600
88.480   467.498
89.480   1034.423
mode 64.000
80.480   200.519
81.480   206.713
82.480   214.551
83.480   224.784
84.480   238.704
85.480   258.734
86.480   290.015
87.480   345.705
88.480   472.586
89.480   1048.509
mode 65.000
80.480   201.356
81.480   207.646
82.480   215.606
83.480   225.999
84.480   240.137
85.480   260.480
86.480   292.250
87.480   348.810
88.480   477.673
89.480   1062.595
mode 66.000
80.480   202.192
81.480   208.579
82.480   216.662
83.480   227.215
84.480   241.570
85.480   262.226
86.480   294.485
87.480   351.914
88.480   482.760
89.480   1076.681
mode 67.000
80.480   203.028
81.480   209.512
82.480   217.717
```

```
83.480    228.430
84.480    243.003
85.480    263.972
86.480    296.719
87.480    355.019
88.480    487.848
89.480    1090.767
mode 68.000
80.480    203.864
81.480    210.445
82.480    218.773
83.480    229.645
84.480    244.436
85.480    265.718
86.480    298.954
87.480    358.124
88.480    492.935
89.480    1104.853
mode 69.000
80.480    204.701
81.480    211.378
82.480    219.828
83.480    230.861
84.480    245.869
85.480    267.464
86.480    301.188
87.480    361.229
88.480    498.022
89.480    1118.939
mode 70.000
80.480    205.537
81.480    212.311
82.480    220.884
83.480    232.076
84.480    247.302
85.480    269.209
86.480    303.423
87.480    364.333
88.480    503.109
89.480    1133.025
mode 71.000
80.480    206.373
81.480    213.244
82.480    221.939
83.480    233.292
84.480    248.734
85.480    270.955
86.480    305.658
87.480    367.438
88.480    508.197
89.480    1147.111
mode 72.000
80.480    207.209
```

```
81.480   214.177
82.480   222.994
83.480   234.507
84.480   250.167
85.480   272.701
86.480   307.892
87.480   370.543
88.480   513.284
89.480  1161.197
mode 73.000
80.480   208.046
81.480   215.110
82.480   224.050
83.480   235.722
84.480   251.600
85.480   274.447
86.480   310.127
87.480   373.648
88.480   518.371
89.480  1175.283
mode 74.000
80.480   208.882
81.480   216.043
82.480   225.105
83.480   236.938
84.480   253.033
85.480   276.193
86.480   312.361
87.480   376.752
88.480   523.458
89.480  1189.370
mode 75.000
80.480   209.718
81.480   216.976
82.480   226.161
83.480   238.153
84.480   254.466
85.480   277.939
86.480   314.596
87.480   379.857
88.480   528.546
89.480  1203.456
mode 76.000
80.480   210.554
81.480   217.909
82.480   227.216
83.480   239.368
84.480   255.899
85.480   279.684
86.480   316.831
87.480   382.962
88.480   533.633
89.480  1217.542
```

```
mode 77.000
80.480   211.391
81.480   218.842
82.480   228.272
83.480   240.584
84.480   257.332
85.480   281.430
86.480   319.065
87.480   386.067
88.480   538.720
89.480   1231.628
mode 78.000
80.480   212.227
81.480   219.775
82.480   229.327
83.480   241.799
84.480   258.765
85.480   283.176
86.480   321.300
87.480   389.172
88.480   543.808
89.480   1245.714
mode 79.000
80.480   213.063
81.480   220.708
82.480   230.383
83.480   243.015
84.480   260.197
85.480   284.922
86.480   323.534
87.480   392.276
88.480   548.895
89.480   1259.800
mode 80.000
80.480   213.899
81.480   221.641
82.480   231.438
83.480   244.230
84.480   261.630
85.480   286.668
86.480   325.769
87.480   395.381
88.480   553.982
89.480   1273.886
mode 81.000
80.480   214.736
81.480   222.574
82.480   232.494
83.480   245.445
84.480   263.063
85.480   288.414
86.480   328.004
87.480   398.486
```

```
88.480    559.069
89.480   1287.972
mode 82.000
80.480    215.572
81.480    223.507
82.480    233.549
83.480    246.661
84.480    264.496
85.480    290.160
86.480    330.238
87.480    401.591
88.480    564.157
89.480   1302.058
mode 83.000
80.480    216.408
81.480    224.440
82.480    234.605
83.480    247.876
84.480    265.929
85.480    291.905
86.480    332.473
87.480    404.695
88.480    569.244
89.480   1316.144
mode 84.000
80.480    217.244
81.480    225.373
82.480    235.660
83.480    249.091
84.480    267.362
85.480    293.651
86.480    334.708
87.480    407.800
88.480    574.331
89.480   1330.230
mode 85.000
80.480    218.081
81.480    226.306
82.480    236.716
83.480    250.307
84.480    268.795
85.480    295.397
86.480    336.942
87.480    410.905
88.480    579.419
89.480   1344.316
mode 86.000
80.480    218.917
81.480    227.239
82.480    237.771
83.480    251.522
84.480    270.228
85.480    297.143
```

```
86.480    339.177
87.480    414.010
88.480    584.506
89.480   1358.402
mode 87.000
80.480    219.753
81.480    228.172
82.480    238.827
83.480    252.738
84.480    271.660
85.480    298.889
86.480    341.411
87.480    417.114
88.480    589.593
89.480   1372.489
mode 88.000
80.480    220.589
81.480    229.105
82.480    239.882
83.480    253.953
84.480    273.093
85.480    300.635
86.480    343.646
87.480    420.219
88.480    594.680
89.480   1386.575
mode 89.000
80.480    221.426
81.480    230.038
82.480    240.938
83.480    255.168
84.480    274.526
85.480    302.381
86.480    345.881
87.480    423.324
88.480    599.768
89.480   1400.661
mode 90.000
80.480    222.262
81.480    230.971
82.480    241.993
83.480    256.384
84.480    275.959
85.480    304.126
86.480    348.115
87.480    426.429
88.480    604.855
89.480   1414.747
mode 91.000
80.480    223.098
81.480    231.904
82.480    243.049
83.480    257.599
```

```
84.480    277.392
85.480    305.872
86.480    350.350
87.480    429.533
88.480    609.942
89.480    1428.833
mode 92.000
80.480    223.934
81.480    232.837
82.480    244.104
83.480    258.814
84.480    278.825
85.480    307.618
86.480    352.584
87.480    432.638
88.480    615.029
89.480    1442.919
mode 93.000
80.480    224.771
81.480    233.770
82.480    245.160
83.480    260.030
84.480    280.258
85.480    309.364
86.480    354.819
87.480    435.743
88.480    620.117
89.480    1457.005
mode 94.000
80.480    225.607
81.480    234.703
82.480    246.215
83.480    261.245
84.480    281.691
85.480    311.110
86.480    357.054
87.480    438.848
88.480    625.204
89.480    1471.091
mode 95.000
80.480    226.443
81.480    235.636
82.480    247.270
83.480    262.461
84.480    283.123
85.480    312.856
86.480    359.288
87.480    441.953
88.480    630.291
89.480    1485.177
mode 96.000
80.480    227.279
81.480    236.569
```

```
82.480   248.326
83.480   263.676
84.480   284.556
85.480   314.601
86.480   361.523
87.480   445.057
88.480   635.379
89.480   1499.263
mode 97.000
80.480   228.115
81.480   237.502
82.480   249.381
83.480   264.891
84.480   285.989
85.480   316.347
86.480   363.758
87.480   448.162
88.480   640.466
89.480   1513.349
mode 98.000
80.480   228.952
81.480   238.435
82.480   250.437
83.480   266.107
84.480   287.422
85.480   318.093
86.480   365.992
87.480   451.267
88.480   645.553
89.480   1527.435
mode 99.000
80.480   229.788
81.480   239.368
82.480   251.492
83.480   267.322
84.480   288.855
85.480   319.839
86.480   368.227
87.480   454.372
88.480   650.640
89.480   1541.521
mode 100.000
80.480   230.624
81.480   240.301
82.480   252.548
83.480   268.537
84.480   290.288
85.480   321.585
86.480   370.461
87.480   457.476
88.480   655.728
89.480   1555.607
mode 101.000
```

```
80.480    231.460
81.480    241.234
82.480    253.603
83.480    269.753
84.480    291.721
85.480    323.331
86.480    372.696
87.480    460.581
88.480    660.815
89.480   1569.694
mode 102.000
80.480    232.297
81.480    242.167
82.480    254.659
83.480    270.968
84.480    293.154
85.480    325.077
86.480    374.931
87.480    463.686
88.480    665.902
89.480   1583.780
mode 103.000
80.480    233.133
81.480    243.100
82.480    255.714
83.480    272.184
84.480    294.587
85.480    326.822
86.480    377.165
87.480    466.791
88.480    670.990
89.480   1597.866
mode 104.000
80.480    233.969
81.480    244.033
82.480    256.770
83.480    273.399
84.480    296.019
85.480    328.568
86.480    379.400
87.480    469.895
88.480    676.077
89.480   1611.952
mode 105.000
80.480    234.805
81.480    244.967
82.480    257.825
83.480    274.614
84.480    297.452
85.480    330.314
86.480    381.634
87.480    473.000
88.480    681.164
```

```
89.480   1626.038
mode 106.000
80.480   235.642
81.480   245.900
82.480   258.881
83.480   275.830
84.480   298.885
85.480   332.060
86.480   383.869
87.480   476.105
88.480   686.251
89.480   1640.124
mode 107.000
80.480   236.478
81.480   246.833
82.480   259.936
83.480   277.045
84.480   300.318
85.480   333.806
86.480   386.104
87.480   479.210
88.480   691.339
89.480   1654.210
mode 108.000
80.480   237.314
81.480   247.766
82.480   260.992
83.480   278.260
84.480   301.751
85.480   335.552
86.480   388.338
87.480   482.314
88.480   696.426
89.480   1668.296
mode 109.000
80.480   238.150
81.480   248.699
82.480   262.047
83.480   279.476
84.480   303.184
85.480   337.298
86.480   390.573
87.480   485.419
88.480   701.513
89.480   1682.382
mode 110.000
80.480   238.987
81.480   249.632
82.480   263.103
83.480   280.691
84.480   304.617
85.480   339.043
86.480   392.808
```

```
87.480    488.524
88.480    706.600
89.480   1696.468
mode 111.000
80.480    239.823
81.480    250.565
82.480    264.158
83.480    281.907
84.480    306.050
85.480    340.789
86.480    395.042
87.480    491.629
88.480    711.688
89.480   1710.554
mode 112.000
80.480    240.659
81.480    251.498
82.480    265.214
83.480    283.122
84.480    307.482
85.480    342.535
86.480    397.277
87.480    494.734
88.480    716.775
89.480   1724.640
mode 113.000
80.480    241.495
81.480    252.431
82.480    266.269
83.480    284.337
84.480    308.915
85.480    344.281
86.480    399.511
87.480    497.838
88.480    721.862
89.480   1738.726
mode 114.000
80.480    242.332
81.480    253.364
82.480    267.325
83.480    285.553
84.480    310.348
85.480    346.027
86.480    401.746
87.480    500.943
88.480    726.950
89.480   1752.813
mode 115.000
80.480    243.168
81.480    254.297
82.480    268.380
83.480    286.768
84.480    311.781
```

```
85.480   347.773
86.480   403.981
87.480   504.048
88.480   732.037
89.480   1766.899
mode 116.000
80.480   244.004
81.480   255.230
82.480   269.436
83.480   287.983
84.480   313.214
85.480   349.518
86.480   406.215
87.480   507.153
88.480   737.124
89.480   1780.985
mode 117.000
80.480   244.840
81.480   256.163
82.480   270.491
83.480   289.199
84.480   314.647
85.480   351.264
86.480   408.450
87.480   510.257
88.480   742.211
89.480   1795.071
mode 118.000
80.480   245.677
81.480   257.096
82.480   271.546
83.480   290.414
84.480   316.080
85.480   353.010
86.480   410.684
87.480   513.362
88.480   747.299
89.480   1809.157
mode 119.000
80.480   246.513
81.480   258.029
82.480   272.602
83.480   291.630
84.480   317.513
85.480   354.756
86.480   412.919
87.480   516.467
88.480   752.386
89.480   1823.243
mode 120.000
80.480   247.349
81.480   258.962
82.480   273.657
```

```
83.480    292.845
84.480    318.945
85.480    356.502
86.480    415.154
87.480    519.572
88.480    757.473
89.480   1837.329
mode 121.000
80.480    248.185
81.480    259.895
82.480    274.713
83.480    294.060
84.480    320.378
85.480    358.248
86.480    417.388
87.480    522.676
88.480    762.560
89.480   1851.415
mode 122.000
80.480    249.022
81.480    260.828
82.480    275.768
83.480    295.276
84.480    321.811
85.480    359.994
86.480    419.623
87.480    525.781
88.480    767.648
89.480   1865.501
mode 123.000
80.480    249.858
81.480    261.761
82.480    276.824
83.480    296.491
84.480    323.244
85.480    361.739
86.480    421.857
87.480    528.886
88.480    772.735
89.480   1879.587
mode 124.000
80.480    250.694
81.480    262.694
82.480    277.879
83.480    297.706
84.480    324.677
85.480    363.485
86.480    424.092
87.480    531.991
88.480    777.822
89.480   1893.673
mode 125.000
80.480    251.530
```

```
81.480   263.627
82.480   278.935
83.480   298.922
84.480   326.110
85.480   365.231
86.480   426.327
87.480   535.095
88.480   782.910
89.480  1907.759
mode 126.000
80.480   252.366
81.480   264.560
82.480   279.990
83.480   300.137
84.480   327.543
85.480   366.977
86.480   428.561
87.480   538.200
88.480   787.997
89.480  1921.845
mode 127.000
80.480   253.203
81.480   265.493
82.480   281.046
83.480   301.353
84.480   328.976
85.480   368.723
86.480   430.796
87.480   541.305
88.480   793.084
89.480  1935.932
mode 128.000
80.480   254.039
81.480   266.426
82.480   282.101
83.480   302.568
84.480   330.408
85.480   370.469
86.480   433.031
87.480   544.410
88.480   798.171
89.480  1950.018
mode 129.000
80.480   254.875
81.480   267.359
82.480   283.157
83.480   303.783
84.480   331.841
85.480   372.214
86.480   435.265
87.480   547.514
88.480   803.259
89.480  1964.104
```

mode 130.000
80.480   255.711
81.480   268.292
82.480   284.212
83.480   304.999
84.480   333.274
85.480   373.960
86.480   437.500
87.480   550.619
88.480   808.346
89.480   1978.190
mode 131.000
80.480   256.548
81.480   269.225
82.480   285.268
83.480   306.214
84.480   334.707
85.480   375.706
86.480   439.734
87.480   553.724
88.480   813.433
89.480   1992.276
mode 132.000
80.480   257.384
81.480   270.158
82.480   286.323
83.480   307.429
84.480   336.140
85.480   377.452
86.480   441.969
87.480   556.829
88.480   818.521
89.480   2006.362
mode 133.000
80.480   258.220
81.480   271.091
82.480   287.379
83.480   308.645
84.480   337.573
85.480   379.198
86.480   444.204
87.480   559.934
88.480   823.608
89.480   2020.448
mode 134.000
80.480   259.056
81.480   272.024
82.480   288.434
83.480   309.860
84.480   339.006
85.480   380.944
86.480   446.438
87.480   563.038

```
88.480   828.695
89.480   2034.534
mode 135.000
80.480   259.893
81.480   272.957
82.480   289.490
83.480   311.076
84.480   340.439
85.480   382.690
86.480   448.673
87.480   566.143
88.480   833.782
89.480   2048.620
mode 136.000
80.480   260.729
81.480   273.890
82.480   290.545
83.480   312.291
84.480   341.872
85.480   384.435
86.480   450.907
87.480   569.248
88.480   838.870
89.480   2062.706
mode 137.000
80.480   261.565
81.480   274.823
82.480   291.601
83.480   313.506
84.480   343.304
85.480   386.181
86.480   453.142
87.480   572.353
88.480   843.957
89.480   2076.792
mode 138.000
80.480   262.401
81.480   275.756
82.480   292.656
83.480   314.722
84.480   344.737
85.480   387.927
86.480   455.377
87.480   575.457
88.480   849.044
89.480   2090.878
mode 139.000
80.480   263.238
81.480   276.689
82.480   293.712
83.480   315.937
84.480   346.170
85.480   389.673
```

```
86.480    457.611
87.480    578.562
88.480    854.131
89.480    2104.964
mode 140.000
80.480    264.074
81.480    277.622
82.480    294.767
83.480    317.152
84.480    347.603
85.480    391.419
86.480    459.846
87.480    581.667
88.480    859.219
89.480    2119.050
mode 141.000
80.480    264.910
81.480    278.555
82.480    295.822
83.480    318.368
84.480    349.036
85.480    393.165
86.480    462.081
87.480    584.772
88.480    864.306
89.480    2133.137
mode 142.000
80.480    265.746
81.480    279.488
82.480    296.878
83.480    319.583
84.480    350.469
85.480    394.911
86.480    464.315
87.480    587.876
88.480    869.393
89.480    2147.223
mode 143.000
80.480    266.583
81.480    280.421
82.480    297.933
83.480    320.799
84.480    351.902
85.480    396.656
86.480    466.550
87.480    590.981
88.480    874.481
89.480    2161.309
mode 144.000
80.480    267.419
81.480    281.354
82.480    298.989
83.480    322.014
```

```
84.480    353.335
85.480    398.402
86.480    468.784
87.480    594.086
88.480    879.568
89.480   2175.395
mode 145.000
80.480    268.255
81.480    282.287
82.480    300.044
83.480    323.229
84.480    354.767
85.480    400.148
86.480    471.019
87.480    597.191
88.480    884.655
89.480   2189.481
mode 146.000
80.480    269.091
81.480    283.220
82.480    301.100
83.480    324.445
84.480    356.200
85.480    401.894
86.480    473.254
87.480    600.295
88.480    889.742
89.480   2203.567
mode 147.000
80.480    269.928
81.480    284.153
82.480    302.155
83.480    325.660
84.480    357.633
85.480    403.640
86.480    475.488
87.480    603.400
88.480    894.830
89.480   2217.653
mode 148.000
80.480    270.764
81.480    285.086
82.480    303.211
83.480    326.875
84.480    359.066
85.480    405.386
86.480    477.723
87.480    606.505
88.480    899.917
89.480   2231.739
mode 149.000
80.480    271.600
81.480    286.019
```

```
82.480    304.266
83.480    328.091
84.480    360.499
85.480    407.131
86.480    479.957
87.480    609.610
88.480    905.004
89.480   2245.825
mode 150.000
80.480    272.436
81.480    286.952
82.480    305.322
83.480    329.306
84.480    361.932
85.480    408.877
86.480    482.192
87.480    612.715
88.480    910.092
89.480   2259.911
mode 151.000
80.480    273.273
81.480    287.885
82.480    306.377
83.480    330.522
84.480    363.365
85.480    410.623
86.480    484.427
87.480    615.819
88.480    915.179
89.480   2273.997
mode 152.000
80.480    274.109
81.480    288.818
82.480    307.433
83.480    331.737
84.480    364.798
85.480    412.369
86.480    486.661
87.480    618.924
88.480    920.266
89.480   2288.083
mode 153.000
80.480    274.945
81.480    289.751
82.480    308.488
83.480    332.952
84.480    366.230
85.480    414.115
86.480    488.896
87.480    622.029
88.480    925.353
89.480   2302.169
mode 154.000
```

```
80.480    275.781
81.480    290.684
82.480    309.544
83.480    334.168
84.480    367.663
85.480    415.861
86.480    491.131
87.480    625.134
88.480    930.441
89.480   2316.256
mode 155.000
80.480    276.618
81.480    291.617
82.480    310.599
83.480    335.383
84.480    369.096
85.480    417.607
86.480    493.365
87.480    628.238
88.480    935.528
89.480   2330.342
mode 156.000
80.480    277.454
81.480    292.550
82.480    311.655
83.480    336.598
84.480    370.529
85.480    419.352
86.480    495.600
87.480    631.343
88.480    940.615
89.480   2344.428
mode 157.000
80.480    278.290
81.480    293.483
82.480    312.710
83.480    337.814
84.480    371.962
85.480    421.098
86.480    497.834
87.480    634.448
88.480    945.702
89.480   2358.514
mode 158.000
80.480    279.126
81.480    294.416
82.480    313.766
83.480    339.029
84.480    373.395
85.480    422.844
86.480    500.069
87.480    637.553
88.480    950.790
```

```
89.480   2372.600
mode 159.000
80.480   279.962
81.480   295.349
82.480   314.821
83.480   340.245
84.480   374.828
85.480   424.590
86.480   502.304
87.480   640.657
88.480   955.877
89.480   2386.686
mode 160.000
80.480   280.799
81.480   296.282
82.480   315.877
83.480   341.460
84.480   376.261
85.480   426.336
86.480   504.538
87.480   643.762
88.480   960.964
89.480   2400.772
mode 161.000
80.480   281.635
81.480   297.215
82.480   316.932
83.480   342.675
84.480   377.693
85.480   428.082
86.480   506.773
87.480   646.867
88.480   966.052
89.480   2414.858
mode 162.000
80.480   282.471
81.480   298.148
82.480   317.988
83.480   343.891
84.480   379.126
85.480   429.827
86.480   509.007
87.480   649.972
88.480   971.139
89.480   2428.944
mode 163.000
80.480   283.307
81.480   299.081
82.480   319.043
83.480   345.106
84.480   380.559
85.480   431.573
86.480   511.242
```

```
87.480    653.076
88.480    976.226
89.480   2443.030
mode 164.000
80.480    284.144
81.480    300.014
82.480    320.098
83.480    346.321
84.480    381.992
85.480    433.319
86.480    513.477
87.480    656.181
88.480    981.313
89.480   2457.116
mode 165.000
80.480    284.980
81.480    300.947
82.480    321.154
83.480    347.537
84.480    383.425
85.480    435.065
86.480    515.711
87.480    659.286
88.480    986.401
89.480   2471.202
mode 166.000
80.480    285.816
81.480    301.880
82.480    322.209
83.480    348.752
84.480    384.858
85.480    436.811
86.480    517.946
87.480    662.391
88.480    991.488
89.480   2485.288
mode 167.000
80.480    286.652
81.480    302.813
82.480    323.265
83.480    349.968
84.480    386.291
85.480    438.557
86.480    520.181
87.480    665.495
88.480    996.575
89.480   2499.375
mode 168.000
80.480    287.489
81.480    303.746
82.480    324.320
83.480    351.183
84.480    387.724
```

```
85.480    440.303
86.480    522.415
87.480    668.600
88.480    1001.663
89.480    2513.461
mode 169.000
80.480    288.325
81.480    304.679
82.480    325.376
83.480    352.398
84.480    389.157
85.480    442.048
86.480    524.650
87.480    671.705
88.480    1006.750
89.480    2527.547
mode 170.000
80.480    289.161
81.480    305.612
82.480    326.431
83.480    353.614
84.480    390.589
85.480    443.794
86.480    526.884
87.480    674.810
88.480    1011.837
89.480    2541.633
mode 171.000
80.480    289.997
81.480    306.545
82.480    327.487
83.480    354.829
84.480    392.022
85.480    445.540
86.480    529.119
87.480    677.915
88.480    1016.924

99.480    2555.719
mode 172.000
80.480    290.834
81.480    307.478
82.480    328.542
83.480    356.044
84.480    393.455
85.480    447.286
86.480    531.354
87.480    681.019
88.480    1022.012
89.480    2569.805
mode 173.000
80.480    291.670
81.480    308.411
```

```
82.480   329.598
83.480   357.260
84.480   394.888
85.480   449.032
86.480   533.588
87.480   684.124
88.480   1027.099
89.480   2583.891
mode 174.000
80.480   292.506
81.480   309.345
82.480   330.653
83.480   358.475
84.480   396.321
85.480   450.778
86.480   535.823
87.480   687.229
88.480   1032.186
89.480   2597.977
mode 175.000
80.480   293.342
81.480   310.278
82.480   331.709
83.480   359.690
84.480   397.754
85.480   452.524
86.480   538.057
87.480   690.334
88.480   1037.273
89.480   2612.063
```

Therefore, wvgdes computes the results for a total of 175 possible supported TE modes of this dielectric slab optical waveguide. In the above listing, the values for mode 0.00 can be ignored; for any other mode, the first column represents the possible propagation angle values, with the maximum possible value of 90 degree. The second column represents the computed values of $\frac{d}{\lambda}$ as explained in Sect. 2.11.

The computed values for modes 1, 2, 3, 4, 5, 6, 7, 8, and 8 are shown in the mode chart in Fig. 3.1. It has been generated with the common ready-to-use and freely available graphics package *gnuplot* [2].

3.6 Estimating Coupling Efficiency Loss from Optical Fiber Misalignment

The ready-to-use C computer language [1] executable **wvgdccl** computes the efficiency loss due to misalignment of two optical fibers, especially when the gap between these is too large, or the fiber axes are offset, or the axes of the two fibers are offset at an angle. The loss estimate calculations are based on the assumption that the

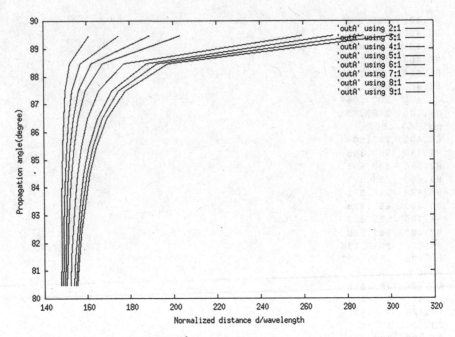

Fig. 3.1 Mode chart for modes 1, 2, 3, and 4. The lowest order mode is leftmost. The normalized waveguide thickness is $\frac{d}{\lambda}$

name profile is Gaussian, which in turn implies that the beam has a *waist*, i.e., minimum diameter. Typing `./wvgdccl` at the Linux shell command prompt generates the help information:

```
$ ./wvgdccl
incorrect|insufficient arguments
batch|command line argument input
./wvgdccl b|B|c|C 1|2|3
<angle offset (degree) > 0.0 ONLY for angle offset fiber junction
<beam waist 1 micron>
<beam waist 2 micron - 0.0 ONLY for inline offset AND angle offset fiber
junction>
<gap length micron > 0.0 ONLY for gap coupled fiber junction>
<offset micron > 0.0 ONLY for offset coupled fiber junction>
<refractive index>
<frequency THz > 0.0 ONLY for gap angle offset fiber junction>
sample batch|command line argument inputi - gap coupled fiber junction
./wvgdccl b|B|c|C 1 0.0 15 0.0 1.5 0.0 1.48 5.0
sample batch|command line argument input axis offset coupled fiber
junction
./wvgdccl b|B|c|C 2 0.0 2 2.45 0.0 10.0 1.48 0.0
sample batch|command line argument input angle offset coupled fiber
junction
```

```
./wvgdccl b|B|c|C 3 5.0 2 1.75 0.0 0.0 1.48 10.0
Gaussian beam profile assumed in all three calculations
```

Using the supplied sample command line input for the three cases mentioned before, the results are listed below.

Gap between fibers:

```
$ ./wvgdccl b 1 0.0 15 0.0 1.5 0.0 1.48 5.0
inline gap coupled fiber efficiency 1.000
```

Offset between fiber axes:

```
$ ./wvgdccl B 2 0.0 2 2.45 0.0 10.0 1.48 0.0
inline axial offset coupled fiber efficiency 0.578
```

Fiber axes offset at a very small angle:

```
$ ./wvgdccl c 3 5.0 2 1.75 0.0 0.0 1.48 10.0
angle gap coupled fiber efficiency 0.982
```

3.7   Estimation of Coupling Ratio of Fused Optical Fiber Directional Coupler (Sect. 2.15.4)

The ready-to-use C computer language [1] executable **wvgddc** implements the simple scheme explained in detail in Sect. 2.15.4 to estimate the coupling ratio of a fused optical fiber directional coupler. Typing **./wvgddc** at the Linux shell command prompt generates the help information:

```
$ ./wvgddc
incorrect|insufficient arguments
batch|command line argument mode
./wvgddc b|B|c|C
<unstretched optical fiber overlap length mm>
<instretched fiber radius micron>
<signal wavelength nanometer>
<test max. coupling ratio>
<test increment value for iteration>
sample batch|command line argument input
./wvgddc b 6 63 1550 0.85 0.25
```

Using the supplied sample command line input as $ **./wvgddc B 6 63 1550 0.85 0.25 > out** redirects the output into a simple text file named "out." The contents of this text output file are shown in Fig 3.2, generated using *gnuplot* [2]:

The periodic nature of the coupling ratio is clear from this graph.

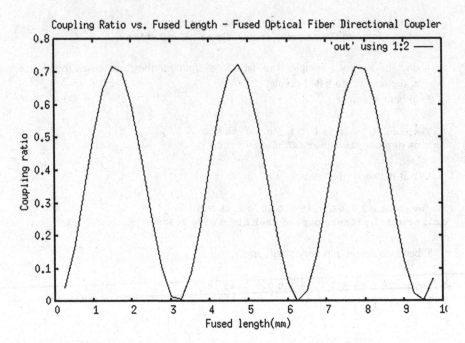

Fig 3.2 Coupling ratio of a fused optical fiber directional coupler

3.8   Estimating Grating Coupler Parameters for Codirectional and Contra-directional Propagation (Sect. 2.15.5)

The ready-to-use C computer language [1] executable **wvgdgrc** computes the grating coupler properties for both codirectional (modes propagating in the same direction) and contra-directional propagation (modes propagating in opposite directions). These are, respectively, the grating period and reflection coefficient, for co/contra-directional propagation. Typing `./wvgdgrc` at the Linux shell command prompt generates the help information:

```
$ ./wvgdgrc
incorrect|insufficient arguments
batch|command line arguments
./wvgdgrc b|B|c|C
<frequency THz>
<refractive index core|guide
<refractive index cladding|substrate>
sample batch|command line argument input
./wvgdgrc c 2 500 4 0.0 1.51 1.50 0.125
```

Using the supplied sample command line argument input generates the following result:

```
./wvgdgrc c 2 500 4 0.0 1.51 1.50 0.125
co-directional propagation
longitudinal propagation constants 1.580467e+07   1.570000e+07
grating period 6.000000e-05 m
contra-directional propagation
reflection coefficient per grating period 0.9925
```

3.9 Estimating Total Energy Loss and Intermodal Dispersion in an Optical Fiber

The ready-to-use C computer language [1] executable **wvgdisploss** inputs the total energy loss in a real-world optical fiber link, and also the dispersion delay amongst two highest order modes in the same waveguide. Typing **./wvgdisploss** at the Linux shell command prompt generates the help information:

```
$ ./wvgdisploss
incorrect|insufficient arguments
batch|command line argument input format
./wvgdisploss b|B|c|C
<operating frequency THz>
<total length of optical fiber km>
<loss per connector dB>
<loss per splice dB>
<loss cable dB/km>
<system loss dB>
<optical fiber core radius micron>
<optical fiber core center refractive index>
<optical fiber cladding refractive index>
<number of connectors>
<number of splices>
<loss type 1|2>
<graded refractive index core profile parameter>
loss type 1 -> attenuation
loss type 2 -> inter modal dispersion
sample command line argument input attenuation
./wvgdisploss b 500 20 0.5 0.1 0.5 3 40 1.5 1.45 2 4 1 0
sample command line argument input inter modal dispersion - step index
multimode
./wvgdisploss c 550 20 0.5 0.1 0.5 3 40 1.5 1.45 2 4 2 0
inter modal dispersion for graded index
multimode optical fiber requires
experimental estimation of variation of
core|cladding refractive indices with wavelength
```

Using the supplied sample command line inot generates the following results.

```
$ ./wvgdisploss b 500 20 0.5 0.1 0.5 3 40 1.5 1.45 2 4 1 0
total attenuation 14.40000 dB
$ ./wvgdisploss c 550 20 0.5 0.1 0.5 3 40 1.5 1.45 2 4 2 0
inter modal dispersion 1.646941e-10 sec/m
waveguide dispersion group delay for fundamental 1.000000e-04 sec for
20.000 km
```

Exercises

1. What useful information can be extracted from the results generated for the case of the graded index fiber of example 3.1? Why is the maximum possible number of supported modes almost twice that of the single-mode fiber examined in the same example?
2. Why do the first few allowed modes for the graded index optical fiber in example 3.1 have a single digit, instead of the familiar (l, m) notation for a step index fiber?

Use **wvgdseigeneng** to determine the allowed transverse wavenumber for TM odd mode propagation in a dielectric slab waveguide with core, cladding dielectric constants, and operating frequency of your choice.

For each of the slab dielectric waveguide design examples in this chapter, compute the longitudinal wave numbers.

References

1. The all-time classic book on C computer language, written by the creators: https://www.amazon.in/Programming-Language-Prentice-Hall-Software/dp/0131103628
2. http://www.gnuplot.info/

Correction to: Optical Waveguides Analysis and Design

Amal Banerjee

Correction to:
Chapters 2 & 3 in: A. Banerjee, *Optical Waveguides*
Analysis and Design,
https://doi.org/10.1007/978-3-030-93631-0

The original versions of chapters 2 and 3 were inadvertently published without online extra supplementary materials (ESM). The chapters have now been updated with online ESM and approved by the author.

The updated original version of these chapters can be found at
https://doi.org/10.1007/978-3-030-93631-0_2
https://doi.org/10.1007/978-3-030-93631-0_3

Appendix A: List of Ready-to-Use C Computer Language Executables

wvgdcsi	Computes the longitudinal propagation constant and related parameters for a step index optical fiber. Uses weak guidance approximation.
wvgdesof	Enables the design of both graded index and step index optical fibers by computing key parameters as V parameter, maximum possible number of modes, core radius for single-mode operation, and so on. Uses the weak guidance approximation.
wvgdseigeneng	Computes the longitudinal propagation constants of slab optical waveguides, for even/odd TE/TM modes, but does not compute mode number.
wvgdseigenA	Computes the longitudinal propagation constants for slab optical waveguide, for both even/odd TE/TM modes and the corresponding mode numbers.
wvgdes	Computes the physical dimensions of slab optical waveguides and mode charts.
wvgdccl	Computes the losses in energy when light is coupled between two optical fibers. Such losses arise if the gap between the two fibers is large, or the axes are offset, or the axes are offset at a small angle.
wvgddc	Computes the coupling ratio of an optical waveguide directional coupler.
wgdgrc	Computes the grating period and coupling ratio of a grating coupler.

© The Editor(s) (if applicable) and The Author(s), under exclusive license to
Springer Nature Switzerland AG 2022
A. Banerjee, *Optical Waveguides Analysis and Design*,
https://doi.org/10.1007/978-3-030-93631-0

Two sets of the ready-to-use C computer language executables, one each for the popular Linux and Windows operating systems, are supplied. The executables for the Windows operating system can be used only under the MinGW (Minimalist GNU for Windows) framework. Appendix B provides a list of URLs that contain detailed step-by-step instructions to download and install MinGW on any Windows operating system computer.

Appendix B: How to Download and Install MinGW on Windows Operating System Computers?

The following Web sites contain detailed step-by-step instructions with screenshots about how to download and install MinGW on Windows operating system machines:

https://www.ics.uci.edu/~pattis/common/handouts/mingweclipse/mingw.html
http://www.codebind.com/cprogramming/install-mingw-windows-10-gcc/
https://www3.ntu.edu.sg/home/ehchua/programming/howto/Cygwin_HowTo.html
https://azrael.digipen.edu/~mmead/www/public/mingw/
https://cpp.tutorials24x7.com/blog/how-to-install-mingw-on-windows

© The Editor(s) (if applicable) and The Author(s), under exclusive license to
Springer Nature Switzerland AG 2022
A. Banerjee, *Optical Waveguides Analysis and Design*,
https://doi.org/10.1007/978-3-030-93631-0

Index

A. Banerjee, *Optical Waveguides Analysis and Design*,
https://doi.org/10.1007/978-3-030-93631-0

Printed in the United States
by Baker & Taylor Publisher Services